Essays on Biocatastrophe

and the
Collapse of Global Consumer Society

Ephraim Tinkham

Engine Company No. 9

Radscan-Chemfall

Est. 1970

Phenomenology of Biocatastrophe
Publication Series Volume 1

ISBN 10: 0-9846046-0-X
ISBN 13: 978-0-9846046-0-9
Davistown Museum © 2010, 2013

Second edition, second printing with extra essays

Cover photo by The Associated Press

Engine Company No. 9
Radscan-Chemfall
Est. 1970

Disclaimer

Engine Company No. 9 relocated to Maine in 1970. The staff members of Engine Company No. 9 are not members of, affiliated with, or in contact with, any municipal or community fire department in the State of Maine.

This publication is sponsored by

Davistown Museum

Department of Environmental History

Special Publication 69

www.davistownmuseum.org

Pennywheel Press
P.O. Box 144
Hulls Cove, ME 04644

Other publications by Engine Company No. 9

Radscan: Information Sampler on Long-Lived Radionuclides

A Review of Radiological Surveillance Reports of Waste Effluents in Marine Pathways at the Maine Yankee Atomic Power Company at Wiscasset, Maine--- 1970-1984: An Annotated Bibliography

Legacy for Our Children: The Unfunded Costs of Decommissioning the Maine Yankee Atomic Power Station: The Failure to Fund Nuclear Waste Storage and Disposal at the Maine Yankee Atomic Power Station: A Commentary on Violations of the 1982 Nuclear Waste Policy Act and the General Requirements of the Nuclear Regulatory Commission for Decommissioning Nuclear Facilities

Patterns of Noncompliance: The Nuclear Regulatory Commission and the Maine Yankee Atomic Power Company: Generic and Site-Specific Deficiencies in Radiological Surveillance Programs

RADNET: Nuclear Information on the Internet: General Introduction; Definitions and Conversion Factors; Biologically Significant Radionuclides; Radiation Protection Guidelines

RADNET: Anthropogenic Radioactivity: Plume Pulse Pathways, Baseline Data and Dietary Intake

RADNET: Anthropogenic Radioactivity: Chernobyl Fallout Data: 1986 – 2001

RADNET: Anthropogenic Radioactivity: Major Plume Source Points

Integrated Data Base for 1992: U.S. Spent Fuel and Radioactive Waste Inventories, Projections, and Characteristics: Reprinted from October 1992 Oak Ridge National Laboratory Report DOE/RW-0006, Rev 8

"Be fruitful and increase, fill the earth and subdue it, rule over the fish in the sea, the birds of heaven, and every living thing that moves upon the earth." (Genesis 1, 28).

Essays on Biocatastrophe

and the

Collapse of Global Consumer Society

The Legacy of Human Ecology

Volume 1 Table of Contents

Preface

This is the first volume in the three volume *Phenomenology of Biocatastrophe* publication series. This publication series explores the biohistory of the imposition of human ecosystems (industrial, social, political, and economic) on or within natural ecosystems. This series is, in essence, the narration of selected stories about what humans are doing to the biosphere. The essays, definitions, databases, editorial opinions, etc. in these texts explore the impact of human activities on the viability of natural ecosystems in our vulnerable, finite, World Commons. No environmental issue is more important than the unfolding tragedy of the depletion of potable water supplies and the contamination of the atmospheric water cycle. The concomitant spectacle of developed, and now aging, western market economies in crisis is intimately connected with the evolving tragedy of the mad rush to oblivion of an out-of-control global consumer society in a biosphere with finite natural resources.

This is the hard copy edition of our frequently updated online series. Suggestions, corrections, and additional information are welcomed. Please send your feedback to the editor at:

CaptainTinkham@gmail.com

The changing text of the *Phenomenology of Biocatastrophe* three volume series may be accessed at:

www.biocalert.org

Printed and Kindle ebook editions of this three volume series may be purchased from:

amazon.com

Printed editions may also be purchased from:

www.davistownmuseum.org/publications.html

I.   Biocatastrophe as a Public Safety Issue

Biocatastrophe is a public safety issue that impacts all nations, communities, families, and individuals without respect to political or religious affiliations or social or economic status. World citizens from all countries and cultures are confronted with a situation of which they may not yet be aware: the ongoing and accelerating depletion of potable water supplies and the contamination of the global atmospheric water cycle (GAWC) by human activities. The degradation of the atmospheric water cycle has one penultimate consequence, the documentation of which is one of several missions of this publication series: the bioaccumulation of environmental chemicals in the maternal cord blood and breast milk of mothers. No greater threat to human health now exists than the invisible spread of the ecotoxins produced by petrochemical man – all of us – into the bodily fluids of the world's mothers and children. The huge databases now published by the US Centers for Disease Control (CDC) – a chemical by chemical analysis of these ecotoxins in human blood (serum concentration) and human urine (creatinine concentrations) – are a graphic demonstration of the consequences of our contamination of the atmospheric water cycle with "environmental chemicals."

The ongoing and accelerating world water crisis is occurring in the context of a multiplicity of other synergistically interrelated ecological crises, including global warming due to greenhouse gas emissions. The diversity of feedback mechanisms that result from global warming insure the looming prospect of cataclysmic climate change. Other environmental crises include loss of ecosystem biodiversity and productivity, mass extinction events, the collapse of oceanic fisheries, and the spread of antibiotic resistant bacteria (ABRB) and other pathogens. The prime movers of these crises include urbanization, deforestation, soil depletion and desertification, and the ubiquitous spread of anthropogenic chemical fallout throughout abiotic and biotic media of the earth's environment. The ongoing Gulf oil spill disaster is a highly visible example of the contamination of the biosphere with a naturally occurring ecotoxin, Louisiana crude oil, a potent primordial symbol of the legacy of petrochemical man. The global financial crisis, triggered by the subprime real estate debacle in the US and an out of control shadow banking network, and exacerbated by unavailability of credit, falling asset values, declining global productivity, and rising debt and unemployment, is much more visible than the seemingly unrelated phenomenon of biocatastrophe. American economists, politicians, academics, and mass media pundits have yet to articulate the connection between the international financial crisis and a looming environmental disaster of which cataclysmic climate change is only one component.

When a brush fire sweeps into the dry grasses of a quiet New England village and threatens outbuildings, barns, and then farm homes and eventually the village center, first responders with their pumper, tanker, and hose lines are the only hope for the community. Humanity is in effect now surrounded by approaching fires, but entirely lacking in the equipment and strategies needed to extinguish them. We now live in a biosphere under siege by forces that cannot be controlled by U.N. conferences on climate change, emergency legislation, governmental financial expenditures, executive fiat, political policy changes, non-governmental organization (NGO) resource allocation, or the adoption of green lifestyles by a growing minority of educated environmentalists and proactive world citizens. The worldwide deterioration of the earth's biosphere is intimately associated with the evolution of a global consumer culture, the rapid growth of megacities in developing countries, and the proliferation of highly productive information, communication, and bioengineering technologies. The obvious benefits of the age of information technology and industrial agriculture obscure the delayed ecological, economic, social, and health physics impact of a rapidly growing world market economy, all of which are discussed in this publication series.

Three issues about a public safety emergency for which there are no first responders demand our attention:

- The identification of the origins of the phenomenon of biocatastrophe – its biohistory – and a sketch of its environmental, health physics, political, psychological, financial, and social components. No component of the phenomenon of biocatastrophe as a biosphere-altering scientific and historical event is more important to all world citizens, regardless of ethnicity or economic status, than the intrusion of manmade ecotoxins into the bodily fluids of women and children.

- The recognition by communities, families, and individuals that there are no governmental entities, economic systems, or political, health care, cultural, or religious institutions that can reverse the consequences of the ongoing deterioration of the earth's natural ecosystems, the accelerating contamination of the atmospheric water cycle, and the consequential eventual collapse of the non-sustainable components of the infrastructure of human ecosystems.

- The determination of the strategies that can be used by individuals, families, and local communities to maintain healthy, sustainable lifestyles in the context of a dying biosphere with finite resources occupied by a predatory global military/industrial/consumer society with an unlimited capacity for environmental pollution and a propensity to deny the consequences of non-sustainable economic activities.

This three volume publication series on the *Phenomenology of Biocatastrophe* focuses on the first two of these issues. Implicit in the unfolding tragedy of our Round-World Commons is the necessity for concerted action by the nations, communities, and

2

individuals (all of us) who are the prime movers of the continuing deterioration of the earth's biosphere. Ultimately, all world communities will be faced with the challenge of maintaining sustainable economies within the next half century in a world of diminishing resources, accelerating chemical fallout, cataclysmic climate change, and a collapsing global economy. Not to debate these issues will only exacerbate the impact of biocatastrophe, impeding the formulation of strategies essential for the survival (but not necessarily the growth) of human civilization.

II.   Biocatastrophe Synopsis

The publication of this survey of the unfolding saga of biocatastrophe is based on the cumulative impact of an interrelated series of scientific facts and historical events:

- Cataclysmic climate change, including global warming, intensified natural weather events, and the prospect of catastrophic sea level rise, is only one component of the larger phenomena of biocatastrophe

- Human population has exceeded the carrying capacity of the Earth's biosphere given the inherent limitations of its natural resources, ecosystems, and food and fresh water production capabilities

- Human activities have generated complex and increasingly diversified and intensifying waves of ecotoxic contaminant pulses as a result of the evolution of a global military/industrial/consumer society and its commodities trading, manufacturing, transportation, weapons production, and communication networks

- The impact of global military/industrial/consumer society is rapidly changing and in fact destroying the biosphere's ecosystem diversity and productivity

- The destruction of natural and human ecosystems is a characteristic activity of pyrotechnic, polymetalic modern society, regardless of the political, religious, social, or economic status or beliefs of the communities involved in these activities

- A fundamental component of biocatastrophe is the evolution and growth of anthropogenic ecosystem infrastructure collapse, initially as invisible third world resource deprivations and epidemics, but also as increasingly visible disruptions of global market economy ecosystems, such as banking and world financial markets, commerce, manufacturing, transportation, education, and health care

- In addition to the ecotoxins that compose chemical fallout, the proliferation of other anthropogenic health hazards influenced by human actions, including antibiotic resistant bacteria, emerging and re-emerging viral epidemics, the spread of toxic pharmaceuticals, and the proliferation of genetically modified organisms are concurrent components of biocatastrophe

- The resulting ongoing annihilation of natural ecosystems, including the contamination of the global atmospheric water cycle, form the context for the probable future collapse of many complex anthropogenic ecosystems

- All of the above phenomena, their synergistic interrelationships, and their cumulative impact on human and natural ecosystems, join together to constitute biocatastrophe, also appropriately called genobiocide, the existence, complexity, and dynamics of which are the subject of intense human denial. This psychological defense mechanism characterizes a large majority of the population of all communities participating in the growth and proliferation of the western model of non-sustainable global military/industrial/consumer society

The following exploration of the impact of human activities on the World Commons of our biosphere, now dominated by human ecosystems that subjugate or destroy natural ecosystems, contains language that many readers may find objectionable. This warning applies in particular to the characterization of the anomalous American consumer culture that evolved after WWII and rapidly spread beyond America's political boundaries to European communities and developing countries with an aspiring middle class. Culture-specific political, social, religious, and psychological traits of a uniquely American ethos have played a major role in exacerbating the human propensity to conquer and control nature, the root cause of biocatastrophe. The description of the small minority of American and world citizens who have had a grossly disproportionate impact in facilitating the rapid degradation of the Earth's biosphere is intended to be contentious. The silver lining in the ongoing and now rapidly unfolding debacle of the non-sustainable western model of a global/military/industrial/consumer society is the recent American Presidential election of 2008. Sixty nine million plus voters said "no" to what has seemed like a growing hegemony, especially in the post-Kennedy era (1963), of a narcissistic American consumer society with ever widening income and health care access disparities described in more explicit terms in the text that follows. A small majority of American voters signaled disapproval of a society where a wealthy minority profits from the growing indebtedness of a middle class with diminishing assets and employment opportunities. The surprising element that accompanied America's changing political milieu in 2008 was the spontaneity and speed of collapsing national and global banking systems and financial markets based on the unsustainable accumulation of debt, a harbinger of the future destiny of global consumer society.

As of the summer of 2010 and the evolution of this text into a three volume series, the initial optimism that characterized the election of the Obama administration is rapidly fading in the context of an intensely partisan Congressional paralysis on health care reform, greenhouse gas reduction, and financial system reform legislation. Also complicating, if not undermining, this optimism, is continuing high unemployment and the emergence of a growing anger in response to Obama administration policies. Accompanying this increasing rancor is the illusion of the end of the "great recession" based on the extraordinary 2009 profits of a still out of control shadow banking network, and the resumption of rapid economic growth in China, Brazil, and, to a lesser extent, in other developing countries while most western market economies linger on the edge of dysfunction. The future ability of America as a nation to stabilize its economy, and thus the political and social viability of its institutions, is unknown. The capacity of American society to effectively adapt to the unfolding age of biocatastrophe is contingent upon its inherent creativity, its capacity for entrepreneurial innovation, the social resiliency of its democratic traditions in an age of increasing stress, and a renewed affirmation of its legacy of justice, fairness, and equal opportunity for all. The fact that successful passage of common sense health care reform occurred on March 21st, 2010 may be a positive indicator that American society still has the capacity to adapt to the social, economic, psychological, political, and environmental stresses of the age of biocatastrophe.

III.   Definition of Biocatastrophe

Biocatastrophe is the simultaneous degradation of the Earth's principal biome ecosystems, including those inhabited by humans, as a result of the radical alteration of the Earth's climate and natural landscapes by a network of nations participating in the highly technometabolic military, industrial, and commercial activities of a global consumer society. The impact of complex western market economies on the World Commons is manifested in the contamination of the biogeochemical cycles of the biosphere with chemical fallout and a concomitant loss of ecosystems biodiversity, productivity, and sustainability. Biocatastrophe is the result of the synergistic impact of the following phenomena:

- Overpopulation and the resulting decline in the affordable, renewable and non-renewable resources necessary to sustain large human populations.

- The systematic exploitation of the Earth's natural resources to create an industrial society that now produces huge quantities of ecotoxins as a component of a global market economy. Anthropogenic and remobilized natural ecotoxins are now circulating throughout all ecosystems, both natural and manmade.

- The ongoing and accelerating contamination of the global atmospheric water cycle by human activities, including warfare, petrochemical production, urban growth, and the proliferation of toxic consumer product and electronic wastes.

- The consequential contamination and/or destruction of food chains and food webs by chemical fallout, which, along with habitat destruction, deforestation, desertification, and soil depletion, are principle factors in the decline of the ecosystem biodiversity and productivity necessary for the survival of human society.

- The historical evolution and worldwide spread of regional, continental, and global warfare that characterizes the age old anthropological quest for control of natural resources and competing cultures.

- The rapid twentieth century increase in the technometabolism of warfare, i.e. the increasing sophistication and power of weaponry and the rapidly accelerating availability of inexpensive mass produced automatic weapons and improvised explosive devices (IEDs). The intense technometabolism of modern warfare is a major source of chemical fallout, global atmospheric water cycle degradation, ecosystem destruction, growing political unrest, and social insecurity.

- The accelerating spread of social unrest and sectarian warfare, facilitated by gross world income and health status disparities, growing food and water stress, and the ready availability of powerful explosives and weapons. Ethnic, religious, political, and class-based social conflicts proliferate and intensify in proportion to declining resource availability – a worldwide Tragedy of the Commons.

- The proliferation of non-sustainable technologies and lifestyles, including the growth of a fossil-fuel-devouring, CO_2 emitting, global consumer culture characterized by a free-trade market economy and high personal, corporate, and governmental indebtedness. The rapidly globalizing world economy is also characterized by the falling incomes of the many consumers and taxpayers in developed nations whose debts funded the profitability of a predatory shadow banking network.

- The concurrent growth and potential collapse of complicated, intricately interwoven, interdependent, vulnerable, anthropogenic ecosystems (world banking, finance and credit markets, manufacturing, transportation, commerce, and trade) that characterize modern, global military/industrial/consumer society.

- The concomitant evolution of cataclysmic climate change, including accelerating global warming, rising sea levels, and intensified natural weather events that result from anthropogenic activities that dramatically alter the water, carbon, oxygen, and nitrogen cycles of the earth's chemosphere and the natural ecosystems upon which humans depend.

- The emergence of new strains of antibiotic resistant bacteria, new forms of viruses, previously unknown pathogens, and the increased proliferation or reemergence of existing disease organisms in an age of global transportation networks.

- The massive proliferation of pharmaceutical chemicals as a result of the advent of advanced medical and bioengineering technologies, the personal care needs of a society of narcissistic consumers, and the worldwide use of growth hormones and antibiotics to increase food production. Contaminant signals from the use and disposal of these pharmaceutical chemicals and growth hormones are now rapidly spreading throughout the biosphere.

- The deliberate or accidental release and dispersion of genetically modified organisms (GMOs), the evolution of pesticide and herbicide-resistant species, the proliferation of vulnerable industrial/agricultural monocultures and the decline of genetically diverse species, all of which ultimately undermine the productivity of the ecosystems of the biosphere necessary to sustain large human populations.

The chemical and biological wastes and pathogens produced by human activity that facilitate the phenomenon of biocatastrophe can be divided into the following eight basic categories.

- ✓ **Anthropogenic radioactivity** produced primarily by military activities (weapons testing) and the generation of nuclear electricity.

- ✓ **Industrial wastes**, including persistent organic pollutants such as pesticides and other ecotoxins produced as a component of military-industrial market economies, industrial agriculture, and hydraulic fracturing for natural gas production.

- ✓ **Consumer product wastes**, including endocrine disrupting chemicals (EDCs) and the proliferation of nanotoxins derived from plastics, electronic equipment, personal care and other consumer products manufactured as a component of a global market economy.

- ✓ **Remobilized naturally-occurring toxins**, including asbestos, cadmium, mercury, lead, arsenic, and other heavy metals produced by military and industrial activities, the burning of coal, oil drilling activities, and consumer product manufacturing.

- ✓ **Greenhouse gases** produced from the combustion of fossil fuels, industrial activities, silicon chip fabrication, the etching of semiconductors, and global-warming-induced permafrost melting.

- ✓ **Antibiotic resistant bacteria (ABRB), HIV/AIDS virus, and other emerging infections** produced as a result of the overuse of antibiotics in animal feed and aquaculture, the over treatment and mistreatment of human health issues, and by the evolution of antimicrobial resistance in hundreds of bacterial strains. HIV is a major ongoing pandemic with the potential to mutate into new forms in combination with drug-resistant strains of TB; other emerging infections are typified by the recent outbreak of a new form of swine influenza, H1N1.

- ✓ **Pharmaceutical wastes** produced primarily as a result of the distribution of drugs used to treat human and animal conditions and the proliferation of growth hormones used to increase food production.

- ✓ **Genetically modified organisms** produced for the purpose of the modification of existing organisms, including food crops, or as the accidental result of the manipulation of the genomes of living organisms.

IV.   Human Ecology: A Bio-Historical Time Line

Ecology is the study of natural systems of living organisms and of their interactions with their environment through the observation and analysis of their life patterns, natural cycles, population changes, and biogeography.

Human ecology is the study of the cultural systems of human society and their interaction with the environment – the observation and analysis of the social behavior of communities, their use of tools and technology, their population and biogeography, and the documentation of their impact on the natural ecosystems of the biosphere.

The ideology of human ecology is the attempt to coordinate human ecosystems with natural ecosystems in the form of sustainable economies, agricultural systems, energy resources, transportation networks, and other non-destructive human activities that will maintain natural ecosystem biodiversity and productivity. The Achilles heel of the ideology of human ecology is its avoidance of the study of the impact of human behavior on the finite natural ecosystems of the biosphere.

The study of human ecology must include the phenomenon of the appropriation of natural resources by human activity, the imposition of anthropogenic (man-made) ecosystems on natural ecosystems, and the evolution of non-sustainable, increasingly complex human ecosystems in natural environments, all of which results in declining biodiversity and ecosystem productivity.

The attempt to reconcile the ideology of human ecology with the historic predatory impact of human activities on natural ecosystems must begin with a panoramic survey of the biohistory of the evolution of human civilization. The study of human ecology must include the anthropological characteristic of using tools for the exploitation of natural resources (i.e. the ecology of technology), the history of world exploration and conquest (i.e. the ecology of warfare), and an analysis of the consequences of the systematic expansion of market economies (i.e. the ecology of money).

The following sketch is biased in its focus on the growth of western market economies and the evolution during the second part of the mature Industrial Revolution of a technologically innovative American empire, which is still the world leader in gross domestic productivity as the Age of Biocatastrophe unfolds.

The Biohistory of Human Civilization

The biohistory of human civilization can be delineated as a series of epochs.

I.   **The Hunting and Gathering Epoch**   **< 6,000 BC**
Found tools were used to create human ecosystems within natural ecosystems. The prime movers of human paleoecosystems were the hand and the foot.

10

II. **The Agrarian Epoch**   **6,000 – 3000 BC**
Ground stone tools and the cultivation of crops expanded human ecosystems as a component of natural ecosystems. The prime movers of agricultural ecosystems were domesticated animals. Human ecosystems began their development within natural ecosystems.

III. **The Epoch of Pyrotechnology and Polymetalism   3000 BC – 1350 AD**
Pyrotechnic, polymetalic human civilization became established as a result of the growth of human ecosystems (political, social, religious, industrial, and economic) within natural ecosystems. The prime movers of polymetalic society were the use of fire to smelt and forge the hand tools and weapons made by bronzesmiths and ironmongers and the use of wind to power the sailing ships of the early empires and raiding communities. The emergence of the distinct cultural ecosystems of human society was characterized by the use of written language to transmit information and maintain power, and the increasing sophistication of weapons production.

IV. **The Epoch of Early Urban Ecosystems**   **1350 - 1750**
The birth and spread of national, regional, and global market economies during this period was characterized by the imposition of human ecosystems on natural ecosystems and the beginning of their systematic destruction by the increasing consumption of natural resources. The prime movers of early urban ecosystems – the proto-Industrial Revolution – were the blast furnace, gunpowder, the printing press, which expanded the availability of written information, and the winds that powered the sailing ships of competing market economies. The evolution and growth of legal, banking, educational, religious, and philosophical systems and theories provided the structures necessary for the spread of increasingly complex human ecosystems.

V. **The Epochs of the Industrial Revolution**   **1750 – 1945**
The Industrial Revolution in its entirety was characterized by the increasing mechanization of human activities and the rapid growth of the exploitation of the natural resources of the biosphere. By the early 19th century, what was the occasional exploitation of natural ecosystems became their ever-expanding destruction and the consequential loss of their biodiversity and productivity. The evolution of the industrial production of machinery, weapons, and consumer products was characterized by the increasing technometabolism of human ecosystems.

Industrial Revolution Part I   **1750 – 1835**

The first stage of the mature Industrial Revolution was characterized by the adaptation of crucible steel production to a wide variety of tools, nautical instruments, and machinery; the use of the steam engine to power textile factories; the invention of the modern reverbatory furnace and rolling mills; and the beginnings of the factory system of mass production of tools with interchangeable parts. The prime movers of the early Industrial Revolution were wood- and coal-fired steam engines, the reverbatory furnace for the production of iron, and the first appearance of machinery used for mass production, typified by Henry Maudslay's block production for the English Navy.

Industrial Revolution Part II   **1835 – 1870**

The second stage of the Industrial Revolution was characterized by the perfection of the factory system of the mass production of guns, tools, and machinery; urbanization; and the increasing exploitation of natural resources, such as coal, for industrial purposes. The prime mover of the second stage of the Industrial Revolution was the steam engine and its application for railroads and steam-powered vessels.

Industrial Revolution Part III   **1870 – 1910**

The third stage of the Industrial Revolution was characterized by bulk steel production and the mechanization of human transportation networks. The steam engine as prime mover was soon joined by the internal combustion engine and the early stages of electronic communications – the telegraph and the telephone. Petrochemical man made his first appearance. Oil joined coal as an important source of fossil fuel energy.

Industrial Revolution Part IV   **1910 – 1945**

The fourth stage of the Industrial Revolution was characterized by increasing mobility, industrialization, manufacturing, and consumer product production, and the growth of the petrochemical industry. The prime movers were the electric power grid and the growing use of the internal combustion engine and its adaptation to aircraft; electronic communications networks now included telephones and radios. A century of uninterrupted global warfare began in 1914, characterized by the rapid increase in the sophistication and power of military weapons.

VI.   **The Age of Chemical Fallout**   **1945 - 1986**

The Age of Chemical Fallout was characterized by the emergence of a rapidly expanding consumer society characterized by a growing middle class, suburbanization, and the maturation of the manufacturing industries that were the basis for the political and economic dominance of the American free enterprise system. The Age of Chemical Fallout was further characterized by radioactive fallout from the advent of the atomic age and the invisible contaminant pulses from the proliferation of endocrine disrupting petrochemical ecotoxins. The prime movers of incipient global military/industrial/consumer society (GMICS) were the accelerated use of the internal combustion engine, nuclear fission, and the invention of microprocessors. Television, satellites, and the rapid spread of computer technology after 1971 expanded global communication networks. The perfection of industrial agriculture resulted in a green revolution that resulted in the rapid growth of world population while also systematically destroying self sufficient local agricultural communities in all nations.

VII. **The Age of Globalization**   **1986 > 2008**

The Age of Globalization was characterized by the worldwide spread of a global consumer society, the design and production of sophisticated computers and other electronic equipment based on new digital and fiber optic technologies (i.e. the birth of the age of information technology,) and rapid advances in bioengineering and pharmaceutical technologies. The legal, banking, and educational systems that provided the structure for the successful functioning of industrial society achieved their maximum productivity. The rapid acceleration in the per acre productivity of the green revolution ended at the same time that vulnerable genetically-modified food crops, such as corn containing the natural bacterial toxin BT, began to dominate world food production. Accelerating contamination of the global atmospheric water cycle (GAWC) and the rapid accumulation of real estate, corporate, consumer, and governmental legacy debts also characterize this epoch.

As the Cold War ended in 1989, global warfare became regionalized as oil wars, sectarian conflicts, ethnic and intertribal warfare, and spreading social unrest facilitated by the increasing availability of powerful explosives and handguns. As a result of globalization, rapidly increasing disparities in income and health care availability occurred in tandem with the demise of regional and local sustainable agriculture. An emerging shadow banking network became the catalyst for the global financial crisis of 2008-? by marketing debt as an asset in the form of collateralized debt obligations (CDOs) and other highly leveraged transactions and issuing insurance for these debts in the form of credit default swaps (CDSs). The prime movers of the Age of Globalization were weapons manufacturers, air transportation networks, the introduction of fiber optic technology, the resulting proliferation of electronic equipment such as computers and cell phones, and the evolution of the worldwide web.

VIII. The Age of Biocatastrophe   > 2008

The Age of Biocatastrophe is characterized by the emerging limitations of the western model of ever-growing market economies, the continuing contamination of the Earth's biosphere by anthropogenic ecotoxins, the accelerating loss of biodiversity and natural ecosystems productivity, the rising threat of genetically modified food crops, the looming menace of terminator genes, and the growing impact of cataclysmic climate change and its multiplicity of feedback mechanisms (loss of albedo, increased methane releases, accelerating glacial melting, rising sea levels). The worldwide financial crisis of 2008-? is a result of the rapid expansion of American and world indebtedness, speculative real estate construction and investments, and non-sustainable consumer product manufacturing since the 1980s. This crisis is a signal of the complexity, interdependency, and financial insecurity of global consumer society and its vulnerability to extra legal manipulation by predatory financial engineers and investors, as manifested in the activities of the World Trade Organization, the growth of private equity funds, and the recent emergence of a vigorous shadow banking network. The recent sudden decline in the gross productivity of the economies of America and other developed nations (but not of China and some other developing nations) is a prelude to the future collapse of the American model of an aging free enterprise system and its probable impact on the viability of sustainable economies, lifestyles, and cultures of all nations. The prime movers of the Age of Biocatastrophe continue to be weapons manufacturers, air transportation networks, electronic equipment, the worldwide spread of the internet, the bioengineers of industrial agriculture, and an American and worldwide shadow banking network. The key development in the Age of Biocatastrophe is that all these prime movers, including the efficient operation of the predatory institutions of finance capitalism, are facilitated by the mother of all prime movers, the worldwide spread of fiber optic technology as a result of the perfection of photonic crystal fibers after 2000. The impact of increasingly intense technometabolism (expenditure of energy) by industrial society is expressed in the growing inventories of anthropogenic ecotoxins, genetically modified organisms, and the financial debt (entropy) produced by the global military/petrochemical/industrial/consumer society network of resource-devouring human ecosystems.

In the mature stage of The Age of Biocatastrophe, the illusion of an ever-growing consumer society characterized by ever-increasing technometabolism is contravened by the limiting factors of finite supplies of natural resources, energy, and money. World and national governments are and will be increasingly unable to ensure that the majority of the world's human populations have access to potable water, adequate food, sustainable employment, education, health care, and social stability. The mantra of "growth" for non-sustainable global consumer society will inevitably give way to the term "survival" as the mandate as well as the challenge confronting national and world economies as they feel the full impact of biocatastrophe, the legacy of human ecology.

Biohistorical Apocalypse

Future generations will suffer the consequences of our egregious insensitivity to the interdependence of the world's ecosystems, and our collective inability to ameliorate the rapidly declining environmental quality of the biosphere we inhabit. Growing social unrest, economic dislocation, and physical and psychological insecurity, all long term consequences of non-sustainable human activities, have also been fostered by the unfortunate evolution of a unique predatory unregulated "free enterprise system" of shadow bankers and hedge fund traders, which flourished in the milieu of American "Reaganomics." These entities used complex stock trading strategies, sophisticated mathematical formulas, powerful computers, and Ponzi schemes to divert the assets and savings of hundreds of millions of the world's working and middle class citizens into the salaries, bonuses, commissions, and profits of a tiny minority of bankers, investors, and profiteers. This diversion of assets was accompanied by the worldwide inflation of real estate, stock, and bond values, the components of a collapsing speculative bubble now threatening the world economy. The increasing control of the world's agricultural and industrial productivity by a decreasing number of ever-growing corporate entities also characterizes this milieu. The ongoing collapse of this unsustainable component of a global market economy in crisis is accompanied by the rapid expansion of drug trading organizations, weapons production and smuggling, piracy, and a wide variety of other organized criminal activities, all of which are characteristic of the early stages of biocatastrophe. The socioeconomic components of the etiology of biocatastrophe and the moral, philosophical, political, and religious values (or lack thereof) expressed in the culture of global consumer society are an integral part of its rapid expansion and inevitable implosion.

A comprehensive description of all of the ecological, scientific, social, and economic components that constitute the phenomenon of biocatastrophe would take a staff of hundreds of individuals, many of them professional scientists with special topic expertise, and would result in a multi-volume encyclopedia of opinion and data on its biogenesis. This publication is, in contrast, an epigrammatic outline of biocatastrophe as a naturally occurring event within a living biosphere that impacts all human communities, written by a few concerned individuals living in isolated New England rural communities.

This handbook is a guide to, as well as a sketch of, the wide variety of components that characterize the phenomena of biocatastrophe in our rapidly changing biosphere during the twilight years of the Industrial Revolution.

After note

The rapid extension of the American subprime mortgage crisis of 2008 into the world financial crisis of 2009 was an unexpected event that occurred during the compilation of this text. Biocatastrophe is envisioned as a process that occurs over a period of centuries, encompassing the evolution of a non-sustainable Industrial Revolution into a post-apocalyptic age of information technology, the challenge of which will be the evolution of a new era of sustainable economies. Will the rapidity of the decline of the non-sustainable component of global consumer society promote a collaboration between the techno-elite engineers of the age of information technology who will be the essential enablers of post-apocalypse sustainable economies, and the workers and laborers (read: electricians, plumbers, teachers, nurses, farmers, truck drivers, etc.) who will make the age of information technology run on time? The deterioration of the financial superstructure of the global market economy is now occurring much more rapidly than we or most other observers anticipated. The economic and social components of the collapse of the American model of a world consumer society and its eerie coincidence with the incipient stages of biocatastrophe have such dramatic consequences for so many world citizens that they obscure the evolving crises of a deteriorating biosphere with finite resources and an increasingly depleted and contaminated atmospheric water cycle. The worldwide humanitarian crisis now unfolding due to the collapsing model of growth-based western market economies serves to marginalize the more invisible phenomenon of environmental degradation, the primary subject of this text, into a Twilight Zone concern. The unfolding saga of the Gulf oil spill disaster may contravene this observation, especially for the fishing and resort communities most affected by this debacle.

V.   The Hotel California: Historical Overview

The term "Hotel California," derived from a popular song of the 1970s by The Eagles, is used in this text as a metaphor for a biosphere that is the ultimate limiting factor for the survival of human society. There is no escape from this hotel and the smorgasbord menu of anthropogenic ecotoxins served in its allegorical dining rooms, the first stop in our journey through its ballroom labyrinths of economic, social, political, technological, and environmental history. The *Phenomenology of Biocatastrophe* publication series is, in essence, a series of essays, synopses, diagrams, observations, op eds, definitions, citations, quotations, and databases encountered on our metaphorical journey through our round-world biosphere. Our confrontation with the evolving tragedy of the World Commons begins in the cafeteria and dining rooms of the Hotel California and quickly moves to the concerts, theaters, picnics, Hyde Park speakers, printed and electronic messages, political signage, research data, and private and public opinions constantly exuded by the seething mass of humanity who occupy our vulnerable Round-World Commons. This publication series is not a carefully peer-reviewed volume of essays by editors at the Oxford University Press, but is, instead, a collection of the epiphanies, messages, and stories encountered in a flat-world journey through a labyrinth of human ecosystems in crisis. The written words of Engine Company No. 9 recapitulate decades of round-world experiences and events, the significance of which may otherwise be lost in the cacophony of biocatastrophe, not to mention in the foggy mists of absentmindedness.

Public Media Commentary

There is now a broad public awareness of the rapid changes occurring in the Earth's biosphere as a result of human activities. American mass media frequently mention the phenomena of global warming. "Carbon footprint" and "greenhouse gases" are now almost as frequent topics of discussion as World Series playoffs or the fluctuating price of gasoline, at least until the advent of the global financial crisis after the demise of Lehman Brothers in September of 2008. Terms such as "global warming" and "global climate crisis" are now joined by "rising sea levels," "melting polar ice," and "permafrost destruction" as acceptable subjects for public media commentary, in part due to the efforts of Al Gore and other public figures to educate the general public about the impact of increasing levels of CO_2 and other greenhouse gases, and the significance of the global climate change that they cause. The mission of this publication series is to foster informed debate on the numerous environmental threats that are not often topics of mass media commentary. A second objective is to explore their relationship to the many political, social, and economic crises that characterize our post-911 world. The recent series of natural disasters, including Hurricane Katrina, the

Haiti earthquake, the Nashville floods, and the Gulf oil spill, help divert public attention from the more invisible challenge of the rapidly diminishing viability of the natural ecosystems we inhabit.

The Hotel California Smorgasbord Labyrinth

When Rachel Carson (1962), and later Theo Colborn, Al Gore, and other environmentalists, first enticed us to explore Hotel California's smorgasbord of currently acceptable environmental topics of discussion, we entered what seemed to be a familiar labyrinth of tired clichés: acid rain, DDT, lead-containing paint, greenhouse gases. In the years following Earth Day 1970, many concerned environmentalists addressed the principle environmental issues of the early years of the Age of Chemical Fallout. As a result of the growing concern about environmental issues in general and chemical fallout in particular, Congress mandated the establishment of the Environmental Protection Agency on Dec. 2, 1970. The consequences of sulfur dioxide in rainfall, DDT in the food chain, and PCBs in Antarctic sea birds were the first in a series of environmental issues that highlighted the importance of the regional and global transport of anthropogenic ecotoxins by the atmospheric water cycle. Major environmental victories were achieved in the reduction or elimination of the production of certain persistent organic pollutants (POPs), sulfur dioxide emissions, and tetraethyl lead in gasoline, yet the diversity and quantity of chemical fallout continues to increase. Since the 1960s, smelly and highly visible human and industrial solid and liquid wastes have been replaced by invisible gaseous plumes of chemical fallout. Selected upscale urban and suburban landscapes may look cleaner than 50 years ago, at least in the more well-to-do neighborhoods of western market economies, but the smorgasbord of anthropogenic ecotoxins has grown exponentially.

The Hotel California: Ecology of Money Rabbit Hole

Immediately encountered in our journey from the dining rooms to the multiple stages, auditoriums, and passageways of the Hotel California are topics of immediate concern to all citizens, especially those pertaining to the economic impact of cataclysmic climate change, and the highly visible loss of habitat for threatened species such as polar bears. Upon leaving the global climate change buffet we encounter the topics of most concern in the lives of everyday citizens – escalating energy, housing, and commodities prices; the reappearance of the spectra of financial uncertainty; unemployment; the sudden unavailability of credit for both small and large businesses; and the social insecurity that results – all subjects of much more immediate concern than melting glaciers. Recent drops in asset values, including real estate and stocks, and the unavailability of employment opportunities for millions of American and world citizens are symptomatic of the rapid unraveling of the global economy. The global

18

financial crisis has close connections with the more invisible genesis of biocatastrophe, a natural consequence of the uncontrolled and unregulated growth of non-sustainable global military/industrial/consumer society. As one penetrates deeper into our Hotel California labyrinth of rapidly changing natural and human ecosystems, the environmental, social, political, and economic topics encountered become more esoteric, obscure, and even unpleasant.

Silent Spring

If we journey into some of the ancient labyrinths of our Hotel California passageways and chambers, to 1962, when Rachel Carson's *Silent Spring* was published, we see the beginnings of the dim outline of the gradual emergence of public awareness of chemical fallout. Even concerned environmentalists had not yet conceived of the possibility of biocatastrophe and the concomitant phenomenon of anthropogenic ecosystems infrastructure collapse (AEIC). Concern about the spread of DDT was followed by awareness of the impact and destructive presence of polychlorinated biphenyls (PCBs), an insulating material found in electrical transformers; Earth Day (April 1970) marked the first time mass media highlighted the importance of these ecotoxins. In the United States and Europe, DDT, PCBs, chlordane, and other ecotoxic chemicals were eventually banned from further production and distribution, but, as long-lived persistent organic pollutants (POPs), these contaminants still persist in the dietary intake of most humans. There are now tens of thousands of biologically significant chemicals to be encountered in the Hotel California's labyrinths of human and natural ecosystems, now characterized by the legacy of anthropogenic petrochemical ecotoxins, pharmaceuticals, and genetically modified organisms.

Ecotoxins and the EPA

The first Earth Day resulted in widespread awareness of the deleterious impact of the supersonic transport plane (SST) on the Earth's protective stratospheric ozone layer. As a result of the probable increase in ultraviolet radiation caused by ozone layer depletion, the US Senate banned the use of the SST by US airlines. During the ensuing decades since its establishment in1970, an increasingly politicized and now demonized Environmental Protection Agency (EPA) has been charged with supervising the cleanup of specific waste disposal (Superfund) sites and documenting the production and release of selected industrial ecotoxins. The EPA *List of Lists* (2006), a database of anthropogenic ecotoxins manufactured by US companies, is a primary source of information about the constituents of ecotoxin contaminant pulses now moving through the environment (see *Appendix C* in *Volume 3*). Enormous annual increases continue to occur in the variety and annual tonnages of worldwide toxic chemical production. The EPA, Centers for Disease Control (CDC), Agency for Toxic Substances and Disease

Registry (ATSDR), and other governmental and non-governmental organizations have the capability to document the amounts of ecotoxins in selected biotic media, including human tissues, breast milk, and maternal cord blood, but have little or no ability to control the size of global contaminant pulses in abiotic and biotic media. In the post-2008 election era, a growing awareness of the extent and significance of an increasingly contaminated global atmospheric water cycle and world food web may result in more detailed studies of the health physics impact of biocatastrophe, but significant world and national governmental mitigation of accelerating genobiocide is unlikely.

Biomonitoring versus Proprietary Information

During the last seven decades, in excess of 80,000 new chemicals have been formulated and produced, the majority by the petrochemical industry; many are discussed or listed in the *Appendices* in *Volume 3*. The alarming reality is that less than 10% of the biologically significant chemicals produced in the last two decades have been evaluated for their toxicity to biotic media, including humans. As one component of our laissez faire "free (but very costly) enterprise system," most of these recently produced chemicals are considered "proprietary." Analysis of their exact chemical composition and, thus, their ecotoxicity, is confidential, subject to the restrictions of privacy designed to prevent misappropriation by competing private enterprise entities. It is in this context that the ability of ATSDR to classify and evaluate ecotoxins has been effectively compromised. As with the shadow banking world of hedge funds and collateral debt obligations, we are living in a petrochemical casino where profit trumps the moral and civic responsibility to protect the World Commons. Its vulnerable atmospheric water cycle is the one limiting factor for the survival of all humans, no matter their socioeconomic status.

Op Ed

The inexorable spread of an increasing variety of ecotoxins is now occurring in the context of the obvious inability of the US and other national and international governments to mobilize the financial and intellectual resources to counteract the rapidly declining viability of modern western consumer culture. The current focus on the unfolding sagas of cataclysmic climate change and the ongoing global financial crisis provides a distraction from the compelling necessity to delineate and debate the fate of human society in a biosphere with finite resources. The current global financial crisis marks a turning point in the history of the market economies of western civilization: we have indulged ourselves in a collective half century spending spree on real estate, rockets to the moon, nonessential consumer products and highly paid predatory financial engineers and media stars. Continued growth of the nonessential components of a global market economy is not a sustainable possibility. Will this reality be acknowledged and debated as the keystone to the growing global financial crisis of 2008-?

VI.   Anthropogenic Ecosystems

Intrusion

The fundamental characteristic of biocatastrophe is the appearance, growth and intrusion of ever-expanding labyrinths of predatory, commodities-consuming, anthropogenic ecosystems into the worldwide network of natural ecosystems and biomes. This complex network of interconnected human ecosystems (human culture in the form of human civilization,) now dominates, and is in the process of obliterating, most other natural ecosystems within the Earth's biosphere. The essential dynamic of biocatastrophe results from the impact of human culture (anthropogenic activities) on the Earth's environment. This impact takes the form of natural resource depletion, loss of species diversity and ecosystem productivity, global climate change, chemical fallout, global and regional warfare, and pandemics, and results in the ongoing and accelerating collapse of natural and anthropogenic ecosystems.

Pyrotechnology and Polymetalism

Two dominant and intertwined characteristics of most human societies are pyrotechnology, the use of fire to make energy and do work; and polymetalism, the use of metals to make tools and weapons used for world exploration, trade, conquest, and global warfare. The modern manifestations of pyrotechnology and polymetalism include the steam engine, internal combustion engine, the production of petrochemicals, electronic communication systems, and nuclear fission. The first polymetalic communities used fire to produce pottery, copper, bronze, and glass. Later industrial polymetalic communities used fire to produce iron, steel, and fossil-fuel-derived petrochemicals, the fundamental components of the infrastructure of modern society. The evolution of the use of the internal combustion engine for transportation systems and agricultural production is the penultimate contribution of petrochemical man to the history of pyrotechnology. The fission process for manufacturing nuclear weapons, electricity and nuclear waste is a unique anthropogenic extension of pyrotechnology. The biogeochemical cycles of the Earth's biosphere provide hemispheric transport of the ecotoxin contaminant pulses produced by these anthropogenic activities.

The Ecology of Tools

A fundamental characteristic of the evolution and growth of human civilization and its many anthropogenic ecosystems is the use of tools to facilitate the efficient functioning of commodities harvesting and exchange networks. No hand tools were more important to the success of world exploration and conquest than the development of high quality weapons (knives, swords, and then firearms of every description). The roots of the global spread of anthropogenic ecotoxins lie in the smelting of metals for the production of the weapons and tools so essential for the success of competing market economies. In turn, one of several goals for victory in warfare, and the principal objective of western market economies, was and is economic gain. The quest for-profit was the fundamental driving force behind an industrial revolution that quickly mechanized the production of tools, weapons, and consumer products.

The Ecology of Money

A fourth fundamental characteristic of human civilization is the evolution of the ecology of money as prime mover of social, political, and community ecosystems. Commodity market economies initially involved the exchange of cultivated food crops and harvested natural resources such as timber, fish, and furs. Increasingly complex human societies soon added precious metals and handmade or forged artifacts to commercial trading networks. The use of minted coinage or printed paper money was eventually accompanied by the emergence of the complex banking system that signaled the birth of regional and global trading empires. Rooted in the Phoenician and Roman trading empires of the early Iron Age, the western model of an ever-growing market economy was born in the mid-16th century at the beginning of the age of exploration and conquest.

The Steam Engine

In the evolution of human culture, western man crossed a Rubicon with the invention of the steam engine. The innovation of boiling water to do work replaced traditional, renewable sources of water, wood, and wind power, paving the way for the replacement of oxen and sailing vessels with railroads and steam ships. The steam engine was the basis for the birth and evolution of modern global military/industrial/consumer society, including global transportation networks, weapons production, and the waging of global warfare. The proliferation of coal-burning steam engines and their resultant production of airborne mercuric sulfides also constituted one of the first examples of the widespread distribution of a remobilized natural ecotoxin by human activity. Increased CO_2 emissions also characterized this second stage of expanding technometabolism (the use of the blast furnace after 1350 is arguably the beginning of the first stage of proto-industrial society).

Pyrotechnology and the Evolution of Biocatastrophe

Biocatastrophe is implicit in the historical evolution of an industrial society that, since the early 19th century, has been characterized by the persistent, systematic, unavoidable discharges of biomass, fossil fuel, and geosphere-derived ecotoxins and climate-altering greenhouse gases. The biosphere, the totality of the Earth's interrelated ecosystems, is the one and only eventual repository for the anthropogenic chemical wastes produced by industrial activities. The evolution of a global market economy with its concomitant transportation, banking, and communications systems is the natural culmination of the growth of an industrial society that has its roots in the appearance of the first pyrotechnic communities. Carbon-dioxide-emitting, charcoal-fired, and then, mercury emitting, coal-fired blast furnaces and steam engines were the first stage of what soon became the rapid proliferation of anthropogenic ecotoxin point sources. Henry Cort's redesigned reverbatory furnace (1784), a key element in the growth of western industrial consumer society, occurred almost two centuries before anyone noticed the ongoing and accelerating contamination of the global atmospheric water cycle and the food webs dependent on this water cycle (Carson 1962; Miller 1970).

Remobilized, Naturally Occurring Ecotoxins

One of the reoccurring phenomena of the growth of human ecosystems is the accidental remobilization of naturally occurring ecotoxins as a component of industrial activity. The ongoing Gulf oil spill is a graphic example of the accidental dispersion of naturally occurring toxins. The Roman use of lead in water pipes and food containers resulted in a significant and well-documented health physics impact; lead continues to be a dangerous component of toys and other consumer products. The burning of coal is a principal source of mercuric sulfide, which is metabolized and remobilized by anaerobic bacteria, creating highly toxic methylmercury. The mining and processing of asbestos into fireproofing materials and other consumer goods have resulted in a late 20th century epidemic of mesothelioma. The use of cadmium in the production of silicon chips for computers, electronic equipment, and photovoltaics produces CdTe (Cadmium telluride), a forgotten external cost of the age of information technology. Many other heavy metals, as ecotoxins of concern, are actively monitored by the CDC and other government agencies (see *Appendix I* in *Volume 3*). The ecotoxic footprints of anthropogenic petrochemicals, including persistent organic pollutants (POPs) and endocrine disrupting chemicals (EDCs) now join the footprints left by remobilized lead, asbestos, cadmium, arsenic, mercury, and other natural ecotoxins.

Radioactive Waste Disposal: An Ethical Paradigm

Commercial nuclear power plants produce large inventories of reactor-derived radioactive wastes. The US Code (USC) requires that the payment for the cost of the disposal of the radioactive wastes generated to create electricity be collected and paid at the time that the electricity is generated. The systematic evasion of waste disposal funding obligations characterizes all past and ongoing nuclear power generation in the United States (Brack 1993). As with the unfunded accumulation of debt for the purpose of facilitating unsustainable economic growth, the unfunded radioactive waste disposal

24

scheme is essentially a series of government-sponsored US Code Title 18 violations. As with other publicly-sponsored debt commitments, future generations of children will inherit the obligation to pay this debt, one of the more invisible legacy costs of a narcissistic consumer culture now spreading to all corners of the world. The legacy of unfunded radioactive waste disposal costs, now an obscure detail of our overall world indebtedness, is symbolic of our collective tendency to use the biosphere as a cesspool for anthropogenic wastes, postponing the costs of their safe disposal for future generations to assume.

The Nuclear Threat

A highly technometabolic global consumer culture with billions of participants may have no other alternative than to accelerate the generation of nuclear energy to counteract the rapid depletion of fossil fuel supplies and the dangers of continued reliance on CO_2-emitting power sources, especially coal. Unfortunately, the hugely expensive costs of constructing new nuclear power plants will greatly exacerbate our collective indebtedness. Western consumer culture already has accumulated huge inventories of "toxic" assets, including real estate, personal, corporate, bank, home, state, and national debt, toxic consumer products, chemical fallout contaminant pulses and newly emerging ABRBs. Continued reliance on aging nuclear power plants, a most toxic asset, is a certainty in view of the looming end of the age of oil and rising CO_2 emissions from fossil fuel combustion. The spread of nuclear weapons technology is a major concern of all national and international governments. The possibilities of future nuclear accidents at aging reactors, terrorist attacks on nuclear facilities, and the proliferation of nuclear weapons, are all components of the phenomenon of biocatastrophe. An unsettling footnote to the issue of nuclear safety is the recent proposal by the Obama administration for the federal government to issue loan guarantees for the construction of two nuclear power plants in Georgia. Unaddressed by President Obama is the question of whether the costs of waste disposal, itself an unsolved puzzle, will be collected at the time the waste is generated as required by federal law, or will they continue to be evaded as is currently the case? (See *Appendices G, H,* and *I* in *Volume 3.*)

Evolution of Global Warfare

Warfare between competing empires is one of the characteristics of the growth of civilizations, the primary source of innovations in metallurgy, and the impetus for the evolution of industrial society. The defeat of the Spanish Armada (1588) and the French/English conflict over control of North America are examples of market economy-based regional warfare that evolved into global warfare in 1914. The global wars of 1914-1918 and 1939-1945 became the global Cold War of 1945-1989, which ended when the Soviet Union collapsed. The Cold War has since evolved into regional oil wars, the spreading sectarian warfare of jihad and its underground terrorist networks, and the proliferation of genocide and tribal conflict on the African continent. In all cases, warfare is a primary cause of habitat destruction and ecotoxin production and distribution. In the later stages of biocatastrophe, regional oil and gas wars between competing market economies will be supplemented by the growing political instability of nonfunctioning governmental entities as the palisaded elite becomes increasingly isolated by widespread social turmoil engendered by unequal or restricted access to potable water, whole foods, and other basic commodities. Lack of employment opportunities and access to public health services and education facilities will exacerbate the impact of malnutrition and result in the evolution of social unrest and the growth of state security organizations. The increasing financial needs of growing state security apparatus will further reduce the availability of scarce public resources.

Anthropogenic Ecosystems

The infrastructure of human communities is an interwoven web of anthropogenic ecosystems: agriculture, manufacturing, commerce, trade, transportation, communications, banking, education, health care, and social services. Radically fluctuating energy and commodities prices, in part due to predatory free market economy speculators, characterize a global market economy with a vulnerable international monetary system, where high indebtedness and loss of asset value characterizes a world banking system in crisis. Other components of the global economy, such as commodities production, transportation, health care, public utilities, and education, are subject to disruption in proportion to the extent and duration of the world financial crisis. The interrelationship between global financial stress due to indebtedness and a biosphere under assault from anthropogenic ecotoxins and exploitation is not yet the subject of mass media analysis. The global financial crisis of 2008-? is a portent for future infrastructure collapse and is closely connected to the high technometabolism of a non-sustainable global economy. The modern global economy, based in part on the accumulation of debt, does not have the capacity or natural resources to extend the conspicuous consumption and lavish lifestyles of the palisaded elite (i.e. the two percent of the world's population with the highest incomes, one hundred thirty five million people) to the masses of working people who are the aspiring participants of a global consumer society that will ideally advance their economic status.

Human Ecosystem Dysfunction

Biocatastrophe is characterized by seemingly unconnected incidents of human ecosystem malfunction. The Chernobyl nuclear accident typifies a unique form of

26

human ecosystem dysfunction, which had vast negative consequences for natural ecosystems. Human ecosystems disrupted by natural weather events include the destruction of large areas of New Orleans after Hurricane Katrina and the widespread disruption of electric utilities in the Galveston and Houston areas after Hurricane Ike. The sudden collapse of the commercial banking system after decades of speculative housing construction, risky corporate loans, and growing consumer indebtedness is yet another example of the cybernetics of biocatastrophe. Regional warfare caused by the need of an oil-devouring American consumer culture to control areas rich in fossil fuel resources is a highly visible contemporary component of the as yet unacknowledged evolution of biocatastrophe. The recent tragedy in Haiti and the ongoing suffering and loss of life of its inhabitants due to lack of shelter is a dramatic illustration of the gap between our collective aid and sympathy for the victims of the Haiti earthquake and our inability to ameliorate their desperate situation. The even more recent unfolding saga of the Gulf oil catastrophe is one more passage in the tragic liturgy of dysfunctional human ecosystems.

The Impact of Human Activity

The future collapse of human ecosystems will begin with local and regional disruptions, as typified by industrial rust belts and abandoned neighborhoods of subprime real estate, accompanied by the systemic loss of natural ecosystem viability and productivity, eventually evolving into sporadic, then sustained, global disruptions of anthropogenic ecosystems such as banking, transportation, trade, and health care. The evolution of modern industrial society is characterized by the hemispheric transport of radioisotopes and chemical fallout manifested in global contaminant pulses in abiotic and biotic media. The potential for global transport of pathogens derived from local and regional epidemics, and pandemics of new and reemerging infectious agents, also characterizes the growing impact of human activity. The invisible presence of anthropogenic chemical and pharmaceutical ecotoxins combines with the age-old human activities of deforestation, resource exploitation and depletion, and warfare, to change the Earth's natural landscape into one dominated by the ecotoxic footprint of human activities.

Need vs. Resources

The mature stage of biocatastrophe will be characterized by the rapid growth of the need for governmental, social, and health care services accompanied by an equally rapid decline in the availability of public resources, including tax revenues and Social Security, retirement, and pension funding. It will also be characterized by a scarcity of adequately trained healthcare and emergency services professionals, and the growing inability of NGOs (non-governmental organizations) to provide services and essential commodities to impacted communities and populations. Biocatastrophe is also characterized by the decreasing size of the affluent communities that can afford to purchase increasingly costly or difficult to obtain commodities, including energy resources, and to maintain access to high quality health care services, whole foods, and fresh water. The phenomena of biocatastrophe is further characterized by the increasing vulnerability of all communities to emerging antibiotic resistant pathogens, GMOs and anthropogenic chemical fallout ecotoxins transported by global biogeochemical cycles, transportation systems, and commodity exchange networks.

Naturally Occurring, Organic Biocatastrophe

Biocatastrophe is the natural, organic (i.e. inherent) result of the synergistic interaction of the social and environmental consequences of anthropogenic activities. Human ecosystems continue to grow in population, engage in regional and local warfare, and produce ever increasing quantities of an ever more widening diversity of ecotoxins. The impact of these activities is manifested in global climate change, loss of natural ecosystem biodiversity and productivity, and the acceleration of the multiplicity of regional and global contaminant pulses that first characterized the age of chemical fallout. The destruction of sustainable human communities is the natural outcome of the continued growth of non-sustainable market economies in the age of global warfare and increasing social unrest. The simultaneous proliferation of multiple industrial activities based on the use of non-renewable energy resources, the spread of global warfare, the development of industrial agriculture, the evolution of a global consumer culture, and the appearance or reappearance of ABRB (antibiotic resistant bacteria), viruses, syndromes, conditions, and epidemics provides the context for biocatastrophe as the natural response of a living biosphere to the unnatural intrusion of resource devouring human ecosystems.

The B Word

Not mentioned as part of modern mass media's commentary on greenhouse emissions and global climate change is the fact that global warming is only one part of a much larger pattern of radical change in the ecology of the biosphere as an integrated whole. The failure to articulate appropriate descriptions of the dynamic changes now occurring in Earth's ecosystems is symbolized by the reluctance of mass media, including such ubiquitous information sources as Cable News Network (CNN), the Public Broadcasting System, *The New York Times, The New York Review of Books*, and other news media, including Wikipedia, to use or define the "B Word" (biocatastrophe). If a concept such as biocatastrophe is not defined in Wikipedia, it must not exist. The reluctance of mass media and the obviously preoccupied Obama administration to articulate and discuss the interconnected series of crises modern society faces, and not

28

just the economic crisis of falling employment and output, is characteristic of the denial of biocatastrophe as an ongoing and accelerating phenomena.

The Tragedy of the World Commons

The Tragedy of the World Commons, the rapid and systematic contamination and destruction of the biosphere, is manifested by the evolution of predatory human ecosystems and their exploitation of natural ecosystems for human purposes. The Tragedy of the World Commons is characterized by the goal of growth-at-any-cost, an ideal rooted in the centuries-old traditions of western market economies, whose early interests in the fish and forests of North America led to the birth of a uniquely sectarian American free enterprise system and the establishment of global/military/industrial/consumer society. The recent sudden collapse of the American commercial and investment banking system, and the global financial crises that followed, are a result of the anthropological propensity of human civilizations to exhaust natural, social, and cultural resources. The recent rapid expansion of American and world indebtedness is a highly visible example of this propensity. The domination and exploitation of natural ecosystems by predatory anthropogenic ecosystems is the natural consequence of the growth and evolution of the unsustainable global market economy of western industrial society.

The Global Consumer Society

Complacency about the growing threat posed by chemical fallout and its cycling within the Earth's biosphere has accompanied the expansion of global trade, the globalization of the world's manufacturing, transportation, and banking industries, and the expansion of worldwide electronic communications, including the internet. The rapid expansion of the global economy has been due in part to the universal appeal of a global consumer society where billions of citizens have the temporary opportunity to begin to enjoy the privileges formerly reserved for a few dozen million American, European, and other well-to-do consumers, circa 1981. The key to the successful functioning of global consumer society is the use of debt to fund ever-expanding production and consumption. Unfortunately, world population and the variety of ecotoxins produced by the global economy are growing as rapidly as the Earth's renewable and non-renewable resources are shrinking. Growing at an even faster rate is the total indebtedness of participants in the western model of consumer society and their real estate, personal, and corporate debts. The limiting factor of money as a nutrient for the growth of consumer society will continue to play a key role in the dynamics of biocatastrophe.

Anthropogenic Ecosystem Infrastructure Collapse

The advent of biocatastrophe is characterized by increasing social unrest and population dislocations caused by competition for, or unavailability of, scarce resources, increasing commodity prices, and widening sectarian, ethnic, and tribal conflicts. The possibility of nuclear attack by nations competing for increasingly costly non-renewable energy resources (e.g. oil wars), and terrorist attacks using nuclear weapons could rapidly accelerate the emergence of full blown biocatastrophe. The mature stage of biocatastrophe will be accompanied by human ecosystem infrastructure collapse, i.e. unexpected disruptions in health care, transportation, education, utility, manufacturing, banking, communications, and consumer product supply systems due to cataclysmic climate change, pandemics, warfare, predatory economic activities, natural resource depletion, workforce unavailability, and/or other unexpected situations. Biocatastrophe triggers could include vaporization of nuclear spent fuel inventories by warfare or terrorist attack, nuclear power plant loss of reactor coolant accidents (LORCAs), pandemics, earthquake and hurricane-derived flashover and washover events, and the collapse of the global market economy due to the accumulation of real estate, consumer, and corporate debt, and unfunded retirement and other legacy obligations.

Synthesis

Biocatastrophe is the natural outcome of the evolution of the age of global military/industrial/consumer society, including the proliferation of industrial agriculture, worldwide environmental degradation, and the ongoing and accelerating loss of the biodiversity of the Earth's vulnerable ecosystems. Biocatastrophe is thus a biogeophysical event with a worldwide social impact, characterized by the hemispheric spread of chemical ecotoxins, the systematic deterioration of the Earth's biosphere, growing regional warfare and social discontent, and the collapse of the infrastructure of urban global consumer society. The relative invisibility of the phenomenon of biocatastrophe is the result of the systematic denial of the natural, organic consequences of the uncontrolled growth of an unsustainable, resource-devouring global military/industrial/consumer society by most (but not all) modern mass media and a large percentage of their many viewers and readers.

VII. Cataclysmic Climate Change

Overpopulation

World Population

Year	Population
1950	2,535,093
1960	3,031,931
1970	3,698,676
1980	4,451,470
1990	5,294,879
2000	6,124,123
2005	6,514,751
2010	6,906,558
2015	7,295,135
2020	7,667,090
2025	8,010,509
2030	8,317,707
2035	8,587,050
2040	8,823,546
2045	9,025,982
2050	9,191,287

Population Division, Dept of Economic and Social Affairs of the UN Secretariat, *World Population Prospects: The 2006 Revision and World Urbanization Prospects.*

A primary cause of biocatastrophe is and will continue to be overpopulation, i.e. the continued growth of total world population beyond the sustainable carrying capacity of the Earth's renewable and non-renewable energy, food, agricultural, oceanic fishery, drinking water, and other natural resources. In 1960, world population had reached three billion people. Numerous texts, including Ehrlich's (1968) *Population Bomb*, appeared warning of the consequences of global population increases. By 2009, world population had exceeded 6.7 billion people. While some of the early predictions of the consequences of population explosion were premature, the combination of increasing human population and the globalization of commerce and trade has resulted in the rapid increase of the per capita consumption of natural resources, nonrenewable energy supplies, and consumer products such as electronic equipment and personal health care products, often made with petrochemical-derived ecotoxins. In a biosphere of limited resources, sustainable populations in excess of eight billion people are inconceivable given the accelerating contamination of the global atmospheric water cycle and the certainty of nonrenewable oil and natural gas shortages after 2020.

Industrial Agriculture

Inherent in the rapid expansion of world population is the challenge of feeding large urban and suburban communities and the growing populations of the rural poor who are no longer self-sufficient in a world of increasingly complex networks of food production. Industrial agricultural systems proliferated in the 20th century due to the increased use of sophisticated mechanized equipment and the ubiquitous application of pesticides. Known as the green revolution, the horse drawn mechanical equipment of the 19th century gave way to complex machinery, agrichemicals, and genetically modified food crops in the last half of the 20th century, facilitating rapid world population growth. Industrial agricultural systems may sustain further population growth for a few decades or more, but a series of looming challenges suggest a mid-21st century limit to their effectiveness. These include the declining availability and increasing cost of fossil fuels and fertilizer, soil degradation and salinization, increasing urbanization and suburbanization, increasing vulnerability of crop monocultures and genetically-modified seed strains to pesticide-resistant pests, proliferation of ecotoxins from chemical fallout, and a number of adverse consequences of global warming and cataclysmic climate change, the latter of which may be the most important limiting factor for the continued viability of industrial agriculture.

Global Warming

The rapid spread of blast furnaces in Western Europe after 1350 resulted in the evolution of proto-industrial societies producing cannons and printing presses for the world-exploring, competing market economies of England, Spain, France, Portugal, and the Netherlands. All had one thing in common, a Celtic metallurgical tradition, which used charcoal-fired furnaces to produce not only the weapons of war or the edge tools of shipbuilders, but the CO_2 emissions that would gradually become one of the most important, if invisible, products of anthropogenic ecosystems. Implicit in the rapidly expanding populations of the 21st century is the expansion of human participation in the western model of global consumer society, assuring accelerating per capita production of CO_2 and other greenhouse gases. The unresolved question of the efficacy of 21st century environmentalism, the greening of global consumer society as it were, is to what extent can accumulating greenhouse gas emissions, especially CO_2 (but not methane) be slowed by governmental policies and lifestyle changes? Unfortunately, only a small component of greenhouse gas emissions is likely to be halted by modifying human behavior and consumer product and transportation vehicle designs.

Al Gore Update on Atmospheric CO_2 Concentrations

On November 12, 2009 Al Gore's Alliance for Climate Protection issued the following statement with respect to the ongoing Congressional climate change legislation: "The concentration of CO_2 in the atmosphere is now 380 parts-per-million (ppm), 100 ppm higher than at the beginning of the Industrial Revolution. And, as a result of the buildup of gases, the temperature is beginning to rise. Adults today have already felt the average global temperature rise more than a full degree Fahrenheit (0.8°C) during our lifetimes. We expect another degree F by 2020 due to past emissions. Based on modeling by an international body of experts studying the climate crisis, the Intergovernmental Panel on Climate Change (IPCC), the temperature could increase by more than 7°F (4°C) by the end of the century in the absence of meaningful efforts to rein in global warming pollution." (Alliance for Climate Protection 2009; http://www.climateprotect.org/climate-challenge/the-facts/). Many American and Europeans may doubt the reality of global warming after the cold weather accompanying the Copenhagen Climate Change Conference, the heavy snows in England, Europe, and America and the Florida freeze of early 2010, but while Minnesota was -40°F, central Maine was +40°F and the temperature in central Greenland was 55°F. Radical changes in weather patterns, including an increase in the intensity of snowfall events in some locations, are characteristic of cataclysmic climate change.

CO_2 Emissions History Time Lines: 10,000 BP – 2010

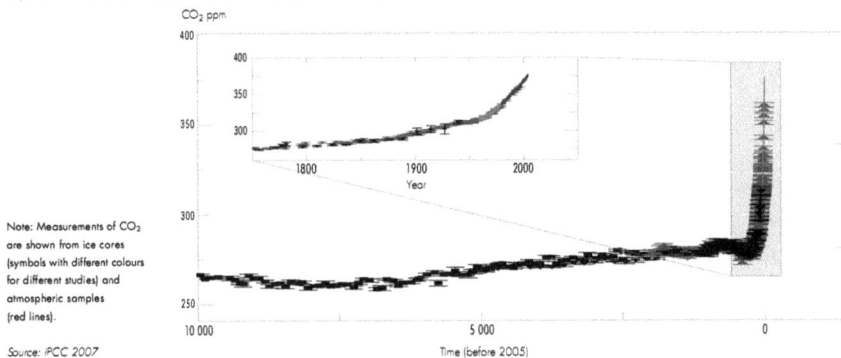

Figure 2.15 Atmospheric concentrations of CO₂ over the last 10 000 years
CO₂ ppm
400
350
300
250
10 000
5 000
0
Time (before 2005)
1800
1900
2000
Year
Note: Measurements of CO₂ are shown from ice cores (symbols with different colours for different studies) and atmospheric samples (red lines).
Source: IPCC 2007

Figure 1 (United Nations Environment Programme 2007).

The United Nations Environment Programme provides this snapshot of the history of CO_2 concentrations in the atmosphere over a period of ten millennia (Figure 1). The sudden rise in atmospheric concentrations of CO_2 has occurred in conjunction with the evolution of industrial society, especially after 1750. More recent data collected by C. David Keeling and the National Oceanic and Atmospheric Administration (NOAA) Earth System Research Laboratory (ESRL) at Mauna Loa, Hawaii, beginning in March 1958 and reproduced in graphic art form by Robert A Rohde from the University of California, Berkeley, provide a compelling illustration of the continuous increase in atmospheric CO_2 concentrations as we enter the age of biocatastrophe (Figure 2).

Global Climate Change

Global warming results from the release of heat reflecting and trapping greenhouse gases by anthropogenic activities such as industrialization, the increasing use of internal combustion engines, the burning of coal, the production of silicon chips and electronic equipment, and the disruption of the carbon cycle by deforestation. Heat which would otherwise escape into space is trapped and reflected back to earth by these gasses. Global climate change is characterized by a wide variety of environmental consequences, which result from global warming and have the potential to significantly impact human ecosystems. The threats from global climate change are the subject of numerous contemporary publications and extensive documentation in scientific media.

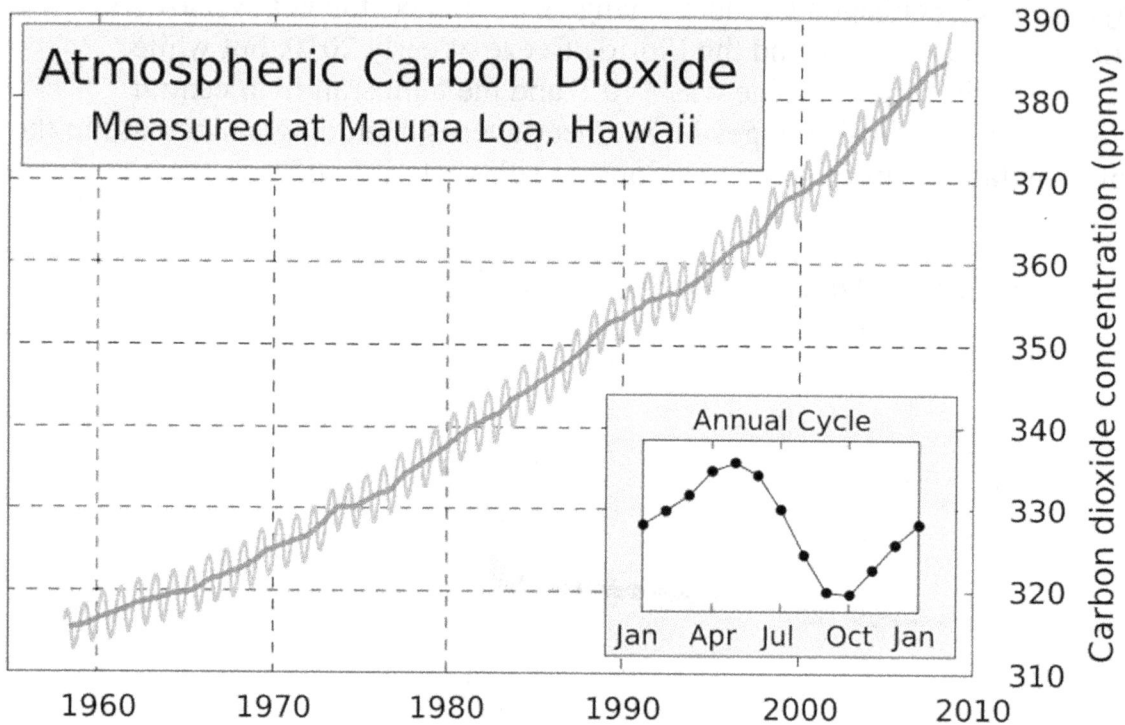

Figure 2 The Keeling curve of atmospheric CO₂ concentrations:
http://en.wikipedia.org/wiki/Carbon_dioxide_in_the_Earth%27s_atmosphere.

The election of the Obama administration signals a new era of more open debate about the consequences of cataclysmic climate change. These consequences include the magnified intensity of natural weather events caused by the increase in total atmospheric water content, melting glaciers and polar ice, rising sea levels, the prospect of radical changes in oceanic ecology, such as coral reef die-off from acidification, and changing currents due to water temperature increases. Cataclysmic climate change, the mature stage of global climate change, will be accelerated by escalating rates of global warming, which are also exacerbated by permafrost melting and methane gas releases, and the increasing absorption of sunlight by the earth's soils and oceans due to decreased albedo produced by polar and Antarctic ice melting. Cataclysmic climate change and the resulting changes in weather and rainfall patterns will further reduce global terrestrial food production capacity. Rising sea levels and the increasing intensity of natural disasters, such as hurricanes and droughts, will result in widespread population dislocations and loss of economic productivity. The many consequences of cataclysmic climate change are now the subject of intensified research and debate by the scientific community, as well as world and national governments already suffering the effects of a sudden and unexpected world financial crisis. The pundits, commentators, economists, academics, and politicians concerned with CO₂ emissions and cataclysmic climate change have yet to connect the dots linking CO_2 emissions to the other forms of chemical fallout.

The Myth of a CO_2 Emissions Cap

An integral component of the denial of the coming age of biocatastrophe is the myth that greenhouse gas reductions, in the simple form of a cap on CO_2 emissions, can mitigate the impact of global warming and sea level rise. Implicit in the media focus on "green" renewable energy strategies is the presumption that a cap on CO_2 emissions, as well as their possible future reduction, will mitigate the ecological consequences of the imposition of human ecosystems on natural ecosystems and solve the problem of global warming. Even if the acceleration of CO_2 emissions could be halted at current levels, total atmospheric CO_2 inventories will continue to rise. Not often mentioned in American media coverage is the huge impact of global deforestation on greenhouse gas absorption, the role of permafrost melting, which releases the greenhouse gas methane, future CO_2 emissions from the world's oceans as a long-term CO_2 storage sink, and the challenge posed by decreased albedo (heat reflection) from rapidly melting polar ice caps. Also not mentioned are the continued emissions of other greenhouse gases, e.g. nitrous oxide from industrial agricultural activities and chlorofluorocarbons, perfluorocarbons, hydrofluorocarbons, tetrafluoromethane, and hexafluoroethane from other industrial activities, including electronic equipment circuit board production.

The Unfortunate Case of Nitrogen Trifluoride (NF_3)

One of the downsides of the age of information technology is the large amounts of energy needed to produce the silicon chips used in digital technology. The production of solar panels, photovoltaic modules, and thin film solar cells also requires a large input of energy, resulting in the net production of greenhouse gasses that actually approach the total produced by gas or coal for similar quantity of solar energy production. A second downside is now emerging with respect to the worldwide production of energy-saving photovoltaic technologies: the concurrent relatively invisible production of nitrogen trifluoride (NF_3) as a byproduct of the plasma etching of silicon wafers used in modern electronic equipment, solar panels, and thin film solar installations. NF_3 is 17,000 times more effective in warming the atmosphere than an equal mass of CO_2. NF_3 is a product of the purification, crystallization, and wafering of silicon for the production of silicon wafers for computers. The recent discovery by the Scripp's Institution of Oceanography at the University of California San Diego that 4,200 metric tons of NF_3 are now in the atmosphere and concentration levels are increasing at 11% per year (Donoghue 2008) reaffirms the necessity of fully evaluating the environmental impact of all emerging "green" technologies. Given the rapid growth of photovoltaic technologies, including thin film solar panels, and the lack of effective gas abatement systems, especially in China, NF_3 production may play a more significant role in cataclysmic climate change than any of the other greenhouse gasses used in the production of semiconductors – trichloroethane (TCA) and trichloroethylene (TCE), or the other orphan greenhouse gasses produced by industrial activity (chlorofluorocarbons, perfluorocarbons, such as the ubiquitous sulfur hexafluoride [SF_6], hydrofluorocarbons, tetrafluoromethane, hexafluoroethane). See *Appendix S* in *Volume 3*, for the International Panel on Climate Control (IPCC) list of the global warming potential of common chemical emissions.

The Irony of Global Cooling

One of the ironic consequences of the Clean Air Act of 1970 and the consequent mitigation of point source air pollution that followed the first Earth Day of April 20, 1970, was that reduction of sulfur dioxide emissions and other visible particulate pollutants had a hidden downside. All such emissions resulted in smog, smoke, or dust pollution of the atmosphere that served to cool the earth's temperature. The creation of unhealthy, unsightly air pollution has been one of the unfortunate footnotes of the Industrial Revolution since the first documented blast furnace began operation in Sweden in the mid-13th century. The ironic downside of the abatement of the "brown" pollutants of industrial activity is that these pollutants no longer served to reflect heat out of the earth's atmosphere, thus eliminating one curb on global warming. Our cleaner atmosphere now more easily absorbs incoming solar radiation as global warming and cataclysmic climate change evolve from walkathons and marathons to a frenetic race to oblivion.

Green Enterprises

A focus of recent and renewed American media interest in global climate change is the potential profit to be made by business interests responding to the now obvious challenges of the growth and worldwide proliferation of global/industrial society and its CO_2-emitting, chemical-fallout-producing, pyrotechnic activities. The expansion of global consumer society provides vast opportunities for-profitable "green" enterprises to ameliorate the insidious environmental impact of climate-altering human activities. With respect to the current worldwide economic crisis, one might view the potential profits to be derived from green enterprises and sustainable economic activities in the age of biocatastrophe as a potential positive component of the ongoing and unfolding economic crisis. The dangers implicit in accelerating greenhouse gas emissions provide a unique opportunity for the techno-elite bioengineers of global consumer culture to formulate innovative solutions to this growing crisis in the free enterprise traditions of inventive adaptation, self-transformation, and economic entrepreneurship. Algae that utilize CO_2 as a food source, algae and other autotrophs that can be converted to biofuels without the deleterious impact of reducing cropland productivity, bioengineered bacteria that can consume plastics and ecotoxins (but what do the bacteria do with them?), and numerous solutions to the challenges of sustainability are all potential contributions of the techno-elite in the age of biocatastrophe. But can green enterprises solve the multiplying and mutating challenges of a rapidly deteriorating world biosphere?

Rising Sea Levels

The most ominous component of cataclysmic climate change is the certainty of rising sea levels. Sea level rise is already occurring at a rate of 3.1 mm/yr (Douglas 1997). Given the increasing rapid melting rate of glaciers and polar ice, this rate will increase. Worst case predictions suggest a catastrophic rise of up to 15 meters by 2200 (Dyson 2008b). Even a one meter rise, a possibility by 2050, will have a devastating impact on low-lying coastal populations and cities throughout the world. Given the rapid construction and protective habitat destruction along world coastlines, the economic and social impact of rising sea levels will be immense. Along with the prospect of a worldwide decrease in food production and the possible acidification of already deteriorating oceanic ecosystems, sea level rise is a subject of extreme concern among world and national governments. Polar bears and other wildlife may not be able to adapt to global warming, melting sea ice, and rising sea levels but humans with their portable ecosystems can adapt to changing climate. Those of us with sufficient financial resources (high technometabolism) can buy our way out of cataclysmic climate change – we hope. Millions of world citizens living in low-lying coastal districts won't have this luxury.

Washover Events

Destruction of large sections of New Orleans and Galveston and the release of huge, but undocumented, inventories of biologically significant ecotoxins into the vulnerable ecosystems of the Gulf of Mexico after Hurricanes Katrina (2005) and Ike (2008) are examples of washover events. Rising sea levels and increasing storm intensities due to elevated atmospheric temperatures and hemispheric water vapor content will be components of future washover events in vulnerable locations such as the Long Island, coastal Florida, Mississippi River flatlands, the Gulf of Mexico coastline, Netherlands, or numerous Asian and African coastal communities, where increasing human populations occupy vulnerable wetlands and lowlands. Hurricane Katrina was America's first nationally televised regional biocatastrophe – an unexpected event that foretold the future. Increasing populations in coastal locations assure higher death tolls and accelerating releases of ecotoxins in future natural disasters.

Cataclysmic Climate Change Mitigation

The possibility exists that the cultivation of genetically engineered carbon-eating plankton, carbon-eating trees, and already available carbon sequestering phytoliths, which entrap carbon in balls of silica in their leaves, can sequester enough CO_2 to halt the acceleration of CO_2 emissions into the atmosphere and possibly even reduce the CO_2 levels of the future to those of the present. Environmental idealists and theoreticians, such as Freeman Dyson (2008), William Nordhaus (2008), and Steven Johnson (2006) postulate rapid advances in biotechnical understanding of the plant genome and the potential for genetically engineered solutions to the long term threat of cataclysmic climate change. Unfortunately, implementation of the solutions to global climate change proposed by the many optimistic techno-elitists is unlikely to be implemented in time to counteract rapidly rising global temperatures. The International Panel on Climate Change (IPCC) predicts a rise of five degrees Celsius above the 18th century global temperature average by 2100. Catastrophic sea level change, altered rainfall patterns, and greatly intensified weather events will already be a major threat at that date. This raises the question as to whether the technical solutions to cataclysmic climate change can be implemented in time to counter the vast future impact of this component of biocatastrophe.

Dissenting Opinion

Not all observers share the widespread opinion that cataclysmic climate change will have a vast impact on human civilization. The brilliant physicist Freeman Dyson, in a recent interview published in *The New York Times Magazine*, expresses an interesting dissenting opinions about the significance and dynamics of global warming. Dyson says the warming "is not global, but local, 'making cold places warmer rather than making hot places hotter'… the carbon may well be salubrious–a sign that 'the climate is actually improving rather than getting worse,' because carbon acts as a ideal fertilizer promoting forest growth and crop yield." Responding to his also well known adversary, James Hanson (NASA), Dyson comments that the carbon dioxide released by burning coal "does not do any substantial harm." Dyson also disagrees with estimates of significant sea level rise and the melting of high level glaciers and asserts that "a warming climate could be forestalling a new ice age." Other notable comments in this interview include "Humans, he says, have a duty to restructure nature for their survival… There's a lot of truth to the statement Greens are people who never had to worry about the grocery bills… The move of the populations of China and India from poverty to middle class prosperity should be the great historic achievement of the century. Without coal it cannot happen… Global warming has become…a party line." (Dawidoff 2009). Dyson, who lives at Princeton, is obviously not in contact with large numbers of the American business community and others who share his view, in contrast to the "party line" espoused in particular by European governments as well as by the relatively tiny numbers of American environmentalists who are actually concerned about the issue. It would be a wonderful development if Dyson's take on cataclysmic climate change trumped the growing evidence of melting glaciers and rising sea levels. He does not, in fact, dispute the scientific accuracy of increasing CO_2 levels in the atmosphere but only questions the significance of its social impact. Unfortunately, the triumph of the Dyson paradigm is extremely unlikely; China and India will be the two most important future contributors to the inevitability of cataclysmic climate change.

The Global Atmospheric Water Cycle

The global atmospheric water cycle is the single most important limiting factor for life on Earth, including the human ecosystems that have intruded themselves within natural ecosystems. Some microorganisms can exist without oxygen, but none without water. Water, along with CO_2 and elemental nutrients, are the main components of living matter formed by photosynthesis. Almost all the living organisms of the biosphere live within a tiny component of the global water supply (Figure 3). Fresh water is only 2.4% of the total global water supply. Of this, only 0.04% is unfrozen surface, riverine, lacustrine, or atmospheric water. All living creatures depend on the relative purity of this fresh (not salt) water. Its vulnerability is illustrated by a startling fact: one gram of water contains 32.4 million quadrillion (3.24×10^{23}) molecules of water. Modern industrial society, with the help of its innovative chemical engineers, has now produced in excess of 100,000 anthropogenic chemicals, including POPs, pharmaceuticals, and a wide variety of other ecotoxins, many of which are highly volatile and/or lipid or water soluble. These are dispersed into the biosphere by pathways ranging from evaporation, incineration, accidental spills, wastewater treatment facilities, to deliberate applications as pesticides, rocket fuel oxidizers, or from the use of personal care products (PCPs) and an infinite array of consumer commodities. These ecotoxins become a part of the global atmospheric water cycle, often one molecule at a time. A rhetorical question for the 21st century: How many different chemical ecotoxins can be incorporated into one gram of water in the global atmospheric water cycle, and then recycled by rainfall events? Documenting the concentration levels of microcontaminants in rainfall in orders of magnitude below current limits of detection (now frequently parts per billion, ppb, or nanograms per liter, ng/L) will be the great future challenge of biomonitoring. Many of these chemicals are present in parts per trillion or parts per quadrillion in the global atmospheric water cycle, with plenty of room in each raindrop for their absorption or adsorption. The most unsettling component of cataclysmic climate change is the invisible ecotoxins transported by the atmospheric water cycle.

VIII. Chemical Fallout

Climate Change and Chemical Fallout

Biocatastrophe begins as gradual local and regional disruptions in, and destruction of, ecosystem biodiversity. Its most well known manifestation is ongoing climate change, currently characterized by increasing levels of atmospheric CO_2 and other greenhouse gases, which are already causing rapidly accelerating global warming and have the potential to cause catastrophic sea levels rises in the future due to polar ice and glacial melting. The most dangerous and most underestimated component of biocatastrophe is not global climate change, but global transport of biologically significant, anthropogenic chemical fallout, including persistent organic pollutants (POPs), ubiquitous endocrine disrupting chemicals (EDCs), methylmercury, and other ecotoxins, which results in the spread and biomagnification of contaminant signals in all abiotic and biotic media, including human tissues, blood, and breast milk.

Chemical Fallout Ecotoxins

Chemical fallout ecotoxins are derived from the petrochemical production of pesticides, a wide variety of consumer products, the burning of fossil fuels, and by other industrial activities of modern pyrotechnic society. These ecotoxins include those produced by internal combustion engines, as rocket fuel derivatives, as pesticides and herbicides, from air transportation contamination of the troposphere, and as components of electronic equipment, photovoltaics, and tens of thousands of consumer products. These ecotoxins are made available and the effects are biomagnified within ecosystems by local, regional, and global biogeochemical transport mechanisms, the most important of which is the global atmospheric water cycle, which connect almost all living organisms in one giant food web. Primary sources of chemical fallout ecotoxins are stationary and mobile fossil-fuel-derived power technologies (steam power, internal combustion engines, and electric power grids), and the toxic radioisotopes and petrochemicals produced for weapons production, agricultural activities, electronic equipment, and consumer products. The extent and complexity of global contaminant pulses of anthropogenic chemical fallout ecotoxins are illustrated by the thousands of chemicals, many of them super toxins, in the Code of Federal Regulations (CFR) Part 40 Section 302.4 (Environmental Protection Agency (EPA) *List of Lists* 2006). See *Appendix C* in *Volume 3*. A small number of the most potent persistent organic pollutants and heavy metals are listed in *Appendix A* in *Volume 3* as monitored and reported by the CDC. This listing includes only a tiny fraction of the biologically significant ecotoxins now a component of the global atmospheric water cycle and thus transported in pathways to human consumption. The EPA *List of Lists* is a further compilation of the reporting levels required for ecotoxins used or produced by American industry.

Toxic Release Inventory (TRI)

The US EPA uses the term toxic release inventory to describe its record keeping of the amounts of extremely hazardous chemicals released by American industry. It is reassuring to know federal government environmentalists are tracking biologically significant chemicals being released to the environment by American industry. There has been some progress in mitigating the size and impact of these releases. There are some flies in the ointment, however. The proprietary status of most chemicals being produced by modern information technology age entrepreneurs is just the tip of our ecotoxic iceberg of ignorance. No international monitoring of the accidental release of toxic chemicals on a worldwide basis is possible as a practical, political, and economic matter. More significant is the total quantity of toxic chemicals produced by GMICS (global military/industrial/consumer society) that are not released to the environment from specific point sources, but incorporated in industrial processes, electronic and other consumer products, industrial agricultural activities, personal care products, transportation systems as fuels and lubricants, or used for other socially useful purposes. If we could quickly fly over all the world's major urban areas as well as all the countryside paved over with shopping malls and suburban communities, especially in the US where so much unused land has been available for development, we would see a huge mass of building materials, asphalt, and concrete, factories and dump sites, all containing large quantities of industrial and consumer products. Buildings and their contents, as well as our landscape, contain large inventories of the same ecotoxins the EPA is trying to track as releases into the environment in the United States. All of these chemicals will be gradually released as demolition, weathering, fire, landfill disposal, leaching, and, especially, deliberate incineration, become the eventual fate of these materials in the next century. Only a tiny percentage of the ecotoxins already manufactured and distributed by our global consumer culture will ever be recovered and properly disposed of. This is the true meaning of toxic release inventory – human society will be calculating these costs as one more toxic asset of our global consumer culture. The faster world population and gross economic productivity grows, the larger the delayed release of our growing inventories anthropogenic ecotoxins becomes.

Waste Electrical and Electronic Equipment (WEEE)

Of particular significance with respect to America's robust production of electronic equipment of every description is the large variety and quantities of ecotoxins that are incorporated in their manufacture. Heavy metals are among the most well known toxic substances in electronic equipment; the enthusiastic salvaging of valuable metals in computers and other equipment by Chinese entrepreneurs has been well publicized (CBS *60 Minutes*) and written about (Puckett 2002). Less well known are the wide variety of other toxins incorporated in the equipment, the most significant of which are the polybromated diphenyl ether (PBDEs) compounds used as fire retardants in electronic equipment and many other consumer products. WEEE is a highly regulated waste stream flow in Europe and the subject of a 2002 directive by the European Parliament and Council. "Separate collection is the precondition to insure specific treatment and recycling of WEEE and is necessary to achieve the chosen level of protection of human health and the environment… Specific treatment for WEEE is indispensable in order to avoid the dispersion of pollutants into the recycled material or the waste stream." (European Parliament and the Council of the European Union 2003). In America, discussion of the wide variety of toxins in consumer products is not yet a subject of extensive public debate. Disposal of highly toxic printed circuit boards, toner cartridges, brominated flame retardants, electronic equipment containing refractory ceramic fibers, and many other WEEE, closely controlled in European communities, is left up to the consumer in America. Only recently have mercury and other highly toxic materials become subject to disposal regulations in selected locations in the United States. The extent of the ongoing dispersion of consumer product ecotoxins is illustrated by the categories of manufactured products subject to controlled disposal in the European community. This listing includes household appliances, IT and telecommunications equipment, consumer and lighting equipment, electrical and electronic tools, toys, leisure and sports equipment, medical devices, monitoring and control instruments, and automatic dispensers, all of which are significant sources of anthropogenic ecotoxins when disposed of in landfills or incinerated in municipal waste facilities. The EPA uses the terminology "nonpoint source (NPS) pollution" to describe the "many diffuse sources" of anthropogenic ecotoxins. Advances are now occurring in recycling strategies in most American landfills, but documentation of the ecotoxic content of most consumer products is infrequent. Many of these ecotoxins are again made available for biogeochemical recycling by careless disposal, reprocessing, or incineration, one more chapter in the saga of the interaction of human and natural ecosystems.

Ecotoxin Bioavailability

Industrial, petrochemical, and consumer product ecotoxins are released to the chemosphere in liquid, gaseous, and particulate form. Most ecotoxins are characterized by having low water solubility and semi-volatile or highly volatile characteristics, or low volatility and high water solubility. In either case, most ecotoxins are, as organic compounds, lipophilic (fat soluble) and are easily transported throughout the food web once incorporated in living organisms. Water soluble ecotoxins can be absorbed within the matrix of just a few water molecules. Likewise, non-water soluble ecotoxins, many of which are highly volatile, and thus, subject to evaporation, can still be captured by the global atmospheric water cycle as adsorbed ecotoxins on the surface of water molecules or airborne particulates. Tropospheric transport of ecotoxins as components of the global atmospheric water cycle were clearly documented as hemispheric in the extent of their distribution as typified by PCBs in Antarctic seabirds, in a now forgotten treatise on chemical fallout (Miller 1970). Such ecotoxins become an important invisible limiting factor for future human population growth due to their health physics impact on both human and natural ecosystems.

Intentionally Released Ecotoxins

Many ecotoxins, as typified by pesticides, herbicides, and pharmaceuticals, are organic chemicals designed to be incorporated into living matter for the purpose of targeting specific pests, either heterotrophs or autotrophs (bacteria, insects, and weeds,) to prevent destruction of valued commodity crops or for enhancing human health. As organic components deliberately targeted to be incorporated within living organisms, these ecotoxins become part of the food chain and are passed back and forth in food webs by the same biogeochemical cycles that transport nutrient elements. As components of the global atmospheric water cycle, they are a frequent presence in rivers, lakes, and streams as well as in surface and public water supplies. They are ubiquitous food chain contaminants as illustrated by research data in *Appendices L, M, and N in Volume 3* as well as a common component of human blood, urine, and lipids, as illustrated in *Appendices A, M, X,* and *Y in Volume 3*. The invention of ecotoxins that quickly decay into harmless metabolites represents a tiny minority of the thousands of petrochemicals and pharmaceuticals manufactured by global military/industrial/consumer society and its multiplicity of biochemists and bioengineers since 1940.

Accidentally Released Ecotoxins

Many ecotoxins are accidentally released into the chemosphere (atmosphere, hydrosphere, and biosphere) as byproducts of industrial activity, transportation systems, and manufacturing processes, as components of personal care products, electronic equipment, or thousands of consumer products. These ecotoxins are organic (carbon-containing) molecules valued for their chemical characteristics as insulators (PCB), fire retardants (PBDE), plastic water pipe components (PVC), plasticizers (bisphenol A), pharmaceuticals (Cialis, growth hormones, antibiotics), cosmetics (musks), grease repellants (perfluorocarbons), rocket fuel additives (perchlorates), smog reducers in gasoline (MTBE), etc. Many others are emitted as accidental byproducts of industrial activity (e.g. methylmercury derived from the burning of coal), refrigeration coolants (chlorofluorocarbons), and the burning of rubbish (dioxins), etc. The global atmospheric water cycle, which links all compartments of the biosphere, including surface dwelling communities of the oceanic hydrosphere (±500 ft), efficiently transports these ecotoxins within all food webs.

Ecotoxin Metabolites and Cogeners

Numerous ecotoxins have a wide variety of chemical metabolites, also called cogeners. The CDC listing of chemicals in *Appendix A* in *Volume 3* graphically illustrates the many "environmental" chemical subtypes, metabolites, and cogeners associated with a specific chemical group. Each ecotoxin subtype, metabolite, or cogener has a different toxicity level, depending on its specific molecular structure and bioavailability during nutrient uptake cycling. Brominated fire retardants (PBDEs), one of many biologically significant groups of ecotoxins not included in the CDC survey, have three principle forms: penta, octa, and deca, according to the number of bromine atoms. These three forms have over 200 cogeners. Many of these cogeners have metabolites whose formation mechanisms in biotic media are not yet understood, but may be more neurotoxic than their primary form. Their modes of transformation into more toxic forms may reflect the impact of sunlight and other chemical ecotoxins acting as catalysts to transform their chemical structure (see *Appendix J* in *Volume 3*).

Synergistic Impact of Multiple Ecotoxins

Multiple ecotoxins in biotic and abiotic media may interact with each other, forming more potent ecotoxins, including ecotoxin metabolites in living organisms. Ecotoxin-ecotoxin synergism, especially within the complex pathways of living organisms, is a hidden component of the proliferation of regional and global contaminant pulses and their growing potency as contaminant signals in biotic media, including pathways to human consumption. Of particular concern is the tendency of polychlorinated biphenyls (PCBs), a particularly commonplace ecotoxin, as well as other halogenated hydrocarbons, to join with other organic chemicals to form new ecotoxins, many of which have not yet been identified. What have been identified as chemicals of concern are multiple combinations of dioxin-furan-PCB cogeners, often the product of combustion, incineration, or cooking (grilling, pan-frying) at relatively high temperatures. The etiology of the growing incidence of immunological, reproductive, and neurological disorders, including autism and behavioral disorders such as ADHD and OCD, is unknown, but may have a link to the proliferation of ecotoxins and ecotoxin cogeners as endocrine disrupting chemicals (EDCs) in the dietary intake of children and the historical increase of the body burdens of these ecotoxins. The many conveniences and benefits of modern technology and the global consumer society make our lives comfortable, with access to advanced health care, educational and global communications systems, and innumerable consumer products, at least for those members of our market economy consumer society who are still employed or independently wealthy. Unfortunately, the many invisible ecotoxins in our daily lives represent the downside of our buy-now, pay-later lifestyles. The rapidly unfolding world financial crisis where total indebtedness exceeds total assets recapitulates our current ecological predicament: we have exploited, depleted, and contaminated our natural resources, fresh waters, and food webs as if the biosphere were a credit card. Few world citizens will escape the consequences of the synergistic impact of multiple ecotoxins.

The Worldwide Threat of Endocrine Disrupting Chemicals (EDCs)

The challenge of evaluating the biological significance of ecotoxin metabolites and cogeners as well as the synergistic impact of multiple ecotoxins is replicated by the difficulty of evaluating the health physics significance of the broad variety of ecotoxins that are endocrine disrupting chemicals. EDCs are omnipresent in consumer products and industrial processes, including industrial agriculture, and range from bisphenol A, phthalates, percolates, and PBDEs in consumer products, such as plastics, computers, cell phones, and furnishings, to POPs, such as DDT and PCBs. The Endocrine Disrupter Exchange (TEDX), referring to the proliferation of EDCs as a component of the worldwide spread of global consumer society, has the following commentary on EDCs.

> In 1991, an international group of experts stated, with confidence, that "Unless the environmental load of synthetic hormone disruptors is abated and controlled, large scale dysfunction at the population level is possible." They could not perceive that within only ten years, a pandemic of endocrine-driven disorders would begin to emerge and increase rapidly across the northern hemisphere. Today, less than two decades later, hardly a family has not been touched by Attention Deficit Hyperactivity Disorder, autism, intelligence and behavioral problems, diabetes, obesity, childhood, pubertal and adult cancers, abnormal genitalia, infertility, Parkinson's or Alzheimer's Diseases. TEDX's findings confirm that each of these disorders could in part be the result of prenatal exposure to chemicals called endocrine disruptors. TEDX has also confirmed that the feed stocks for most endocrine disrupting chemicals are derived from the production of coal, oil, and natural gas... It is clear that endocrine disruption, like climate change, is a spin-off of society's addiction to fossil fuels. Setting aside the effects of endocrine disruptors on infertility, and just considering their influence on intelligence and behavior alone, it is possible that *hormone disruption could pose a more imminent threat to humankind than climate change.* (TEDX 2009; http://www.endocrinedisruption.com/endocrine.fossilfuel.php)

Point Source versus Non-Point Source Pollution

Earth Day 1970 marked the first organized emergence of an environmental movement that was aware of the proliferation of ecotoxins that characterized the age of chemical fallout. This day marked the beginning of an era of governmental attempts to mitigate the impact of chemical fallout; among the first of a series of legislation was the Clean Air Act of 1970 (42 U.S.C. § 7401). This act targeted specific generators and emitters of pollutants, such as sulfur dioxide, chlorinated hydrocarbons, and many of the other chemicals listed in the *Appendices* in *Volume 3*. Among the most notable achievements of the environmental movement was the ban of PCB production in 1976. The closure of the Monsanto chemical company facilities for the production of PCBs is an example of the elimination of a specific point source of an important form of chemical fallout. Numerous other chemicals and ecotoxins, including DDT, Tetra-ethyl lead in gasoline, and sulfur dioxide, have been the subject of successful regulatory action. In terms of the glass half full/half empty metaphor, as applied to regulation of environmental ecotoxins, the numerous specific point sources of biologically significant ecotoxins that were mitigated after 1970, constitute the major achievement of the environmental movement during the last four decades. Unfortunately, the phenomenon of chemical fallout in the biosphere, especially as transported via the global atmospheric water cycle, continues unabated in the 21st century. Part of this is a result of the continued production of ecotoxins in third world countries, especially in the rapidly growing economies of India, China, and other far eastern nations. A second component in the unabated ubiquitous presence of chemical fallout ecotoxins is the introduction of thousands of new variations of POPs (persistent organic pollutants) and entirely new ecotoxins, such as nanotoxins produced by the rapidly expanding network of information technologies. Perhaps the most important component of chemical fallout in the future will be the proliferation of non-point source pollution. Non-point source pollution is best exemplified by the vast expanse of consumer product and industrial debris produced by a global consumer society where the generic invisible emissions of landfills, industrial blight, backyard incinerators, and the less obvious discharges from domestic, commercial, and industrial environments and production facilities cannot be monitored and regulated as can specific "smokestack" point sources. In some cases, pollution sources, such as automobile emissions, can be partially mitigated by legislation and technological innovation, but in the unfolding Age of Biocatastrophe, invisible non-source point pollution generated by an expanding consumer culture and the rapidly growing urban centers of developing nations will constitute the major component of ecotoxin discharges into the biosphere.

Report Units; Limit of Detection (LOD)

Regional and hemispheric transport of ecotoxin contaminant pulses may occur at levels below the LOD (limit of detection), which vary widely among the chemical groups being measured by the CDC, EPA, or other monitoring agency. Reporting units vary in the CDC reports from µg/L (parts per million per liter) of urine or blood for heavy metals to ng/L (parts per billion per liter) for polycyclic aromatic hydrocarbons (Centers for Disease Control and Prevention 2005). See *Appendix A* in *Volume 3*. Some dioxins have a limit of detection measured in the pg/g or trillionths of a gram (lipid analysis,) but the predominant reporting units are ng/g (parts per billion per gram) of lipids in serum and µg/L in urine. See *Volume 2: Biocatastrophe Lexicon, Part III: Commonly used Reporting Units* as well as the CDC definition of limits of detection. Of particular note is the rapidly changing ability to more closely document contaminant signals, as typified in the CDC (2005) report with its lower limits of detection for PCBs, which are now typically 10.5 ng/g maximum, and the reporting levels used in the state of Washington (Department of Ecology 2006) for PBDE, which are pg/g (parts per trillion per gram). Many LODs vary from chemical to chemical and person to person. Even if detectable with current nanotechnologies and picotechnologies, many ecotoxins have not yet been the subject of biological monitoring or scientific study. Ecotoxins in public water supplies are a notable example of contaminants that evade modern surveillance technologies due to their low concentration levels. As these ecotoxins are transferred from one trophic level to the next, their concentrations are biomagnified, as exemplified by DDT (dichlorodiphenyltrichloroethane), which when present in water at levels of 0.00005 parts per million (ppm) becomes magnified in predatory birds at concentrations approaching 30 ppm (Odum 1983).

IX.   The Ecology of Biocatastrophe

The Chemosphere

The migration and distribution of greenhouse gases and toxic chemical fallout occur within a finite, enclosed chemosphere that consists of an integrated atmosphere, hydrosphere, and indwelling biosphere. The constant exchange of nutrient elements within the abiotic and biotic environments of the chemosphere results in increasing exposure of all organisms to a growing inventory of anthropogenic ecotoxins. The worldwide spread of ecotoxins is the result of anthropogenic activities (what humans do within the biosphere); the proliferation of these manmade poisons occurs within a natural chemosphere constituted by fundamental biogeochemical cycles (e.g. water, carbon, oxygen, nitrogen, potassium, etc.), which are the basis for life on earth.

Bioaccumulation by Absorption and Adsorption

Once released into the abiotic environment (air, water, sediments) as organic (carbon containing) compounds, most ecotoxins are absorbed (internalized) or adsorbed (integrated onto outer layers) by inorganic nutrient elements such as potassium, phosphorus, nitrogen, iron, and magnesium or by the water molecules in the global atmospheric water cycle. Airborne particulates subject to atmospheric transport may also absorb or adsorb organic ecotoxins, which are often precipitated by the rainfall events of the global atmospheric water cycle. Organic detrital matter, the residue of living organisms, also may absorb or adsorb ecotoxins from the environment, as well as release them during decay. Bacteria, phytoplankton, zooplankton, fungi, and other single celled organisms universally ingest ecotoxins along with nutrient elements as food, constituting the first stage of the transfer of ecotoxins up the food web. As associated components of nutrient elements and organic matter, ecotoxins are incorporated in and bioaccumulate in the food chain. The primary exposure pathway for these ecotoxins in food chains for humans and other animals is ingestion.

Transport and Biomagnification of Ecotoxins

Ecotoxins, as components of biogeochemical cycles, are incorporated into living protein by micro-consumers (bacteria, fungi), or by the photosynthetic activities of autotrophs (plants and algae), which create most living organisms through the chemical transformation of CO_2, water, and inorganic nutrient elements into living organisms. Once nutrient elements and the organic materials they form are contaminated by absorbed or adsorbed ecotoxins, often on a molecular level, each step in the biogeochemical cycling of nutrient elements as organic molecules also transports the ecotoxins, which also often have an organic (carbon-containing) molecular structure.

Due to the transfer of nutrient elements from one trophic (feeding) level to the next, ecotoxins in living organisms are often biomagnified in each successive trophic level as a component of elemental nutrient cycling. Organic (e.g. POPs) and inorganic ecotoxins (e.g. methylmercury) tend to concentrate at the highest trophic levels of the food web, including predatory game fish (tuna, salmon), predatory birds (eagles, falcons), and in the tissues, blood and breast milk of humans. Pregnant women and newborn children are particularly at risk from biomagnified ecotoxins. Frequently tested indicators of chemical ecotoxins in humans are serum (for ecotoxins in lipids), urine, and blood.

Biogeochemical Cycles

The phenomenon of biocatastrophe is characterized by the biogeochemical cycling of ecotoxins as a series of plumes in both abiotic (nonliving, e.g. water and air) and biotic (living, e.g. plants and animals) media. Toxic chemicals are transported by biogeochemical cycles that are often hemispheric in their nutrient exchange patterns between the abiotic and biotic environments. An ecotoxin, such as diclofenac, which was originally administered to cattle in India, can be recycled in the feces of feral dogs or rats feeding on the carcasses of contaminated cattle, which are then ingested by microorganisms and recycled as micro-contaminants in globally transported food products. Transport time can vary from months to decades. The downside of globalization is the global transport of thousands of ecotoxins as invisible parts per trillion components of numerous foods, fibers, and consumer products inadvertently contaminated with anthropogenic ecotoxins from obscure source points. European biomonitoring agencies use the term "transboundary" to describe the atmospheric cycling of ecotoxins that impact multiple nations. NOAA (www.noaa.gov) has extensive documentation of the trans-Pacific flow of ecotoxins and smoke from China to the United States. The global pulses of ecotoxins, their biogeochemical cycles, and their increasing concentrations in higher trophic levels of the biotic environment are not subject to governmental control by legislation, policy changes, international protocol, or lifestyle changes, our desire to maintain healthy ecosystems or implement sustainable economies notwithstanding.

Bioaccumulation of Ecotoxins in Humans

Mechanisms for the direct bioaccumulation of ecotoxins in humans are well documented. Exposure pathways for ionizing radiation from anthropogenic radioactivity as, for example, the biologically significant isotopes CS_{137} and PU_{239}, depend on the chemical forms of the isotopes. Cesium 137 follows the potassium cycle in nature and when ingested (the primary pathway,) provides a whole body dose. In contrast, the predominant pathway for plutonium 239 exposure is inhalation with ingestion as a secondary pathway. Exposure to PU_{239} occurs primarily in the lung;

ingested PU_{239} is a bone-seeking ecotoxin; the carcinogenic, teratogenic, and mutagenic effects of both isotopes are well documented (United Nations 1972; 1976). Direct exposure to the thousands of biologically significant endocrine disrupting chemicals (EDCs) now present in the biosphere also occurs via the ingestion and inhalation pathways as well as by absorption and the cross-placental transfer of amniotic fluids. Once present in the human body as either water soluble or fat soluble ecotoxins, EDCs, which range from persistent organic pollutants (POPs) such as PCBs, plasticizers (BPA), fire retardants (PBDEs), to oxidizers in explosives (perchlorates) can be either absorbed or adsorbed by cellular tissues. EDCs thus disrupt hormone signaling and alter genetic expression; the rapid increase in EDCs as environmental contaminants in the late 20th century and early 21st century is correlated with the rapid increase in autism spectrum disorders, learning disorders, obesity, type II diabetes, cancer, and many other health physics phenomena.

Natural and Cultural Limiting Factors

The presence of the basic nutrient elements of carbon, oxygen, nitrogen, potassium, and other elemental chemicals are the limiting factors, along with water, sunlight, and temperature, for life on Earth. The interwoven, interdependent biological and cultural ecosystems of global military/industrial/consumer culture are highly vulnerable to contamination and disruption by human activities and anthropogenic ecotoxins. Underlying the efficient functioning of human ecosystems such as health care, education, transportation, consumer product availability, and communications are the fundamental limiting factors of food and fresh water. For human culture and its ecosystems, the availability of uncontaminated basic nutrient elements, clean air and potable water, is subject to disruption in proportion to the increasing complexity of its social organizations. All other cultural limiting factors are dependent on these primary natural limiting factors. Once a basic limiting factor, such as the global atmospheric water cycle, is compromised by depletion and accelerating ecotoxin contamination, human society and its interrelated anthropogenic ecosystems become increasingly stressed.

Global Distribution of the World's Water

Global water distribution summary.

- Oceans 97.5%
- Fresh water 2.5%
    - Glaciers 68.7%

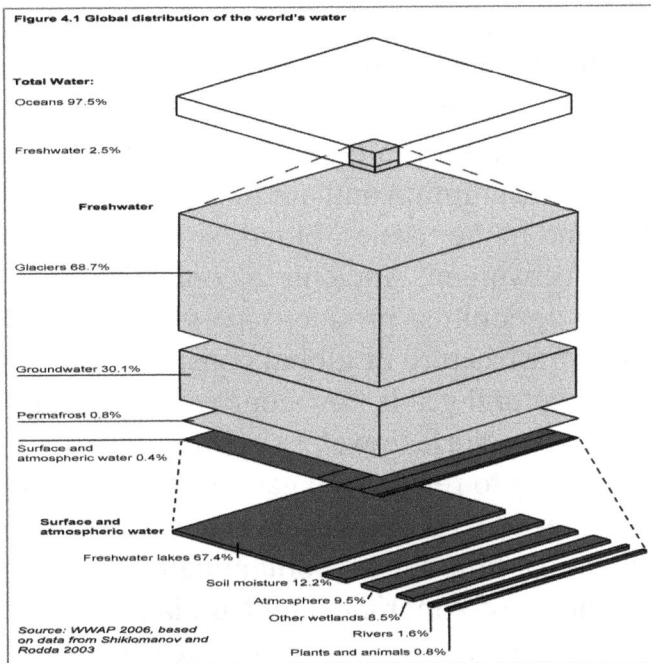

Figure 4.1 Global distribution of the world's water

Total Water:
Oceans 97.5%

Freshwater 2.5%

Freshwater

Glaciers 68.7%

Groundwater 30.1%

Permafrost 0.8%

Surface and atmospheric water 0.4%

Surface and atmospheric water

Freshwater lakes 67.4%

Soil moisture 12.2%
Atmosphere 9.5%
Other wetlands 8.5%
Rivers 1.6%
Plants and animals 0.8%

Source: WWAP 2006, based on data from Shiklomanov and Rodda 2003

- o Ground water 30.1%
- o Permafrost 0.8%
- o Surface and atmospheric water 0.4%
  - Fresh water lakes 67.4%
  - Soil moisture 12.2%
  - Atmospheric water content 9.5%
  - Wetlands 8.5%
  - Rivers 8.6%
  - Living plants and animals 0.8%

Figure 3 Source: United Nations. (2006). *Water a shared responsibility: The United Nations world water development report 2*. World Water Assessment Program. http://www.unesco.org/water/wwap/wwdr/wwdr2/.

Atmospheric Water Cycle Transport Vectors

The sketch of the global distribution of the world's water in 2006 from the UN publication *Global Environment Outlook GEO4* (United Nations Environment Programme 2007, 118) provides the information that fresh water accounts for only 2.5% of the total water in the biosphere. Glaciers account for 68.7% of fresh water, ground water 30.1%, and permafrost 0.8%. The global atmospheric water cycle consisting of surface and atmospheric water constitutes only 0.4% of the 2.5% fresh water share of the global water supply. It is this tiny component of the water in the biosphere that is regularly available for use by living organisms, whether through photosynthesis or biological processes. The UN sketch further breaks down components of the water cycle into fresh water lakes, soil moisture, atmospheric water content, wetlands, and rivers. Only 0.8% (of 0.4% of 2.5%) of surface and atmospheric water is present within living plants and animals. Contamination of the global atmospheric water cycle by anthropogenic ecotoxins has its greatest impact in the context of the biogeochemical uptake of water by the plants and animals of the world's ecosystems living within what is, in essence, one large bowl of soup, i.e. the biosphere. The rapidly expanding world water crisis is a result of the depletion of potable surface water supplies in combination with the proliferation of anthropogenic chemical fallout.

The Limiting Factor of Water I

Uncontaminated, potable, fresh water is the most essential component of healthy ecosystems. The global atmospheric water cycle of the biosphere, including much of the surface waters of the world's ecosystems, is now contaminated with anthropogenic chemical ecotoxins, usually in amounts below our current limits of detection. These ecotoxins bioaccumulate in pathways to human consumption and are universally present in the biotic media, such as birds, fish, and in the tissues, blood, and urine of babies, children, and adults, and the breast milk of women. Ancient, uncontaminated aquifers are being rapidly depleted for industrial agriculture productivity, growing suburban and urban communities and the vast infrastructure of global consumer society. The rapid increase in hydrofracturing to recover natural gas is now contaminating huge quantities of underground water in aquifers as well as surface water supplies due to several hundred chemicals used in the water injected to release the gas trapped in shale rock formations. Underground and surface discharges of pesticides, herbicides, and other ecotoxins and their remobilization as chemical fallout now contaminate all surface water supplies. Underground and surface discharges of American and Russian nuclear wastes, as well as those of other countries, are examples of important but nearly invisible world threats to the future integrity of fresh and saltwater ecosystems. The frequent, often well-documented, discharges of anthropogenic petrochemicals as liquid waste and the continuous ubiquitous nonpoint source discharges of ecotoxins into terrestrial, riverine, and marine environments are now an obvious component of the deterioration of the Earth's water quality, a major threat to human health, and among the most important causes of biocatastrophe. Every rain shower is a chemical fallout event that illustrates the constant recycling of these ecotoxins.

The Limiting Factor of Water II

The unavailability of fresh water for huge Asian urban communities will be the greatest limiting factor for future world population increases and will dwarf the impact of shortages of water in southeastern and southwestern suburban American communities. Himalayan glacial melting, deforestation, and ecotoxin contamination runoff constitute a major limiting factor for future urban Asian population growth. The gradual depletion of North American aquifers and the long term impact of fresh water unavailability will gradually erode the productivity of vast areas of crop lands currently producing the corn, wheat, and soybeans that are essential for feeding much of the world's population. Future population growth in the American southwest will be severely limited by water shortages. Communities with extensive financial resources may be able to establish innovative water supply strategies, such as desalinized seawater. The future of the large underclass of workers and immigrants living in the southwest in a world of acute water shortages and water contamination is unknown. Water pipelines from northern Canada to Arizona are not beyond the realm of possibility but only in a world that can overcome the impact of global financial collapse. In a future world of water stress, the increasing expenses of potable water will be matched by the decreasing ability of individuals and governments to pay its cost. What once was the ready availability of free potable water in many third world communities is now becoming the availability of potable water as a commodity marketed by the corporate entities of a global consumer society. The alternative is the growing health threat of increasingly polluted natural water sources (rivers, wells, surface water sources).

The Limiting Factor of Food

Drought and other increasingly intense weather events, which will result from cataclysmic climate change, have the potential to suddenly impact the world supply of essential agricultural products, especially in vulnerable North American agricultural regions. Urbanization, deforestation, salinization, and soil depletion are worldwide threats to food production capacity. Food shortages as a result of the sudden appearance of pesticide and herbicide-resistant pests in industrial agricultural systems are a future probability. The intrusion of unwanted genetically altered species and their gradual alteration of crop productivity in vulnerable agricultural systems is a further possible limiting factor in world food production. As with fresh water availability, future income disparities and a growing lack of public and NGO resources will exacerbate the difficulties of obtaining increasingly costly healthy food.

Soil and Crop Nutrient Content Depletion

A relatively invisible component of the evolution of biocatastrophe and a key factor in the depletion of food production capacity as a limiting factor for world population is the ongoing worldwide depletion of topsoil. What was an average depth of 21 inches in the 18th century on North American farmlands has been reduced to an average depth of six inches. The impact of industrial agricultural activities has accelerated the depletion of topsoil, which was well underway in the Midwest when the Great Dust Bowl occurred in the 1930s. Sub-Saharan Africa is currently the location of the most rapid and extreme topsoil depletion. Closely associated with topsoil depletion is nutrient content changes in agricultural products, which are even more closely associated with industrial agricultural activities. Radical changes in iron content in spinach have been documented between 1948 (158 mg/100g) to 1965 (28 mg/100g); the last official survey was in 1973 (2.1mg/100g) (Hartmann 2009, 22-3). Losses of mineral content in produce, including calcium, magnesium, and potassium, have been noted between 1963 and 1999 in snap beans, broccoli, carrots, lettuce, peaches, peas, strawberries, and tomatoes, though not in such extreme amounts as the depletion of iron in spinach (Mayer 1997).

Ecotoxins in Food Webs I

In 1970, compelling evidence of the worldwide transport of persistent organic pollutants, specifically PCBs in Antarctic seabirds, was published (Miller 1970). The age of chemical fallout was well underway. As of 2010, global military/industrial/consumer society has manufactured and distributed over 100,000 types of chemicals and pharmaceuticals of every description. Only a few thousand of these have been studied to determine their potential health effects, but a conservative estimate would be that the military, industrial, agricultural, and commercial activities of global consumer society have produced and distributed at least 10,000 biologically significant forms of chemicals. All of the "environmental" chemicals tracked in human urine, blood, and lipids by the CDC and other research organizations (EWG.org) and in wildlife (BioDiversity Research Institute) are also present in food webs as the principal pathway for the uptake of the ecotoxins in these studies.

Ecotoxins in Food Webs II

Many biologically significant ecotoxins are transported in food webs in quantities too small to monitor, i.e. on a molecular level. The CDC has recently been able to observe dioxin-furan ecotoxins emitted primarily from the incineration of plastics and electronic equipment in municipal waste facilities and urban dumpsites in reporting units of trillionths of parts per gram of media (pg/g). Few other ecotoxins are as dangerous as dioxin-furans, but many other unmonitored ecotoxins now contaminate worldwide food webs in concentration ratios of parts per trillion or parts per quadrillion. Some more well known ubiquitous ecotoxins, such as Hg in the form of methylmercury, have been observed in seafood, especially sushi (tuna fish), as well as in birds and mammals with reporting levels above one part per million (1 µg/g) and in lesser concentrations in popular consumer products such as Quaker Oatmeal and Hershey's Chocolate Syrup (350 ppt and 257 ppt respectively) (www.healthobservatory.org/library.cfm?refID+105040) due to Hg contaminant signals in high fructose corn syrup (HFCS). "Hot spots" of methylmercury contamination in New England mammals have now been documented in concentrations approaching parts per ten thousandths (Evers 2007). (See *Appendix O* in *Volume 3*.) Significant worldwide progress has been made in lowering the presence of highly toxic POPs (DDT, aldrin, dieldrin, etc. see *Appendix B* in *Volume 3*), many of which were banned for production or use in Europe, but not in many other countries, by the Stockholm Convention. Their lingering presence and continuing overseas production ensures ongoing bioaccumulation and biomagnification in all biotic media, including human urine, blood, and lipids, illustrating the longevity of chemical contaminants in the environment. These and thousands of other biologically significant ecotoxins are now in world food webs, both as a result of their deliberate production and distribution for socially and economically useful purposes, and as a result of their accidental discharge in soils, landfills, and wastewater or by incineration, evaporation, or demolition. The ongoing, if invisible and difficult to document, contamination of food webs is among the most challenging components of biocatastrophe, lending itself to denial, neglect, and lack of accurate biomonitoring for economic, political, and psychological reasons. Why worry about micro-contaminants in the food chain if they are below the limit of detection, and in modern industrial society their production and dispersion is beyond practical remediation?

Ecosystem Collapse

The proliferation of ecotoxins from industrial activities, transportation systems, electronic equipment, consumer product production, warfare, and agricultural activities exacerbate the intensity and rates of natural ecosystem collapse. Human populations are absolutely dependent on ecosystem productivity and diversity. The decline in the productivity of the ecosystems of the biosphere is a primary limiting factor, along with the availability of water, which will determine human population levels in the centuries to come. The phenomenon of ecosystem collapse includes a decline in the productivity of the world's fisheries and deterioration of world food production capacity, both highly visible results of the loss of biodiversity in aquatic and terrestrial ecosystems. The certainty of the rapid decline in the productivity of the Earth's major food producing ecosystems, and the resulting scarcity and expense of essential commodities, will be the fundamental cause of the social unrest and population dislocations that characterize the physiogenesis of biocatastrophe in the future. Global warming is only one component of the much larger phenomenon of worldwide ecosystem collapse, which is intimately connected with a wide variety of human activities, only one of which is greenhouse gas emissions.

Loss of Biodiversity

Biocatastrophe is characterized by the collapse of the biodiversity that has been created by the evolution of life forms during the last two billion years of the Earth's history. The loss of biodiversity, an evolving mass extinction event, is a result of the impact of anthropogenic activities on the ecosystems that constitute the indwelling biosphere of our finite chemosphere. Habitat destruction is the root cause of the loss of biodiversity and takes many forms. The ongoing loss of biodiversity in all ecosystems is the result of the synergistic interaction of cataclysmic climate change, urbanization, deforestation, salinization, resource depletion, global warfare, and the introduction of genetically modified organisms. The impact of these anthropogenic activities is especially enhanced by chemical fallout ecotoxins produced by the industrial activities of global consumer society. Worldwide loss of ecosystem biodiversity and productivity is the universal result of the impact of human ecosystems on natural ecosystems.

Oceanic Dead Zones

Rapid depletion of oceanic fisheries was already well underway by the mid twentieth century. The rapid decline in the New England cod fisheries (before 1940) was a predictor of the future decline of oceanic fisheries worldwide. Decimation of the Pacific salmon fishery (2008) is only one example of the rapid decline of many fish species (e.g. haddock, halibut, swordfish). The rapid spread of oceanic dead zones due to chemical washout (not only nitrates but many other ecotoxins) and worldwide over-fishing are the basis for the current prediction that oceanic fisheries will no longer supply any significant portion of the world's food supply by 2040 or 2050. Also impacting oceanic fishery viability is the increasing acidification of seawater by CO_2 deposition, a seemingly benign form of chemical fallout that actually poses a major threat to oceanic ecosystems. Methylmercury contaminated tuna fish are the tip of the iceberg of the onslaught of multiple chemical fallout global contaminant pulses in marine food webs. The vast mid-ocean accumulations of floating trash, typified by the great Pacific garbage patch and its ubiquitous high tide line presence, are cultural indicators of the declining health of the world's oceans. The ongoing oil spill in the Gulf of Mexico is a spectacular, highly visible, and very costly, regional oceanic disaster and a prophetic symbol of the entropy inherent in imposing fossil fuel ecosystems on, and within, a natural ecosystem.

Visibility

Biocatastrophe becomes more visible as its social impact increases due to the accelerating depletion of non-renewable and renewable energy and other natural resources, the increasing cost of essential commodities and services, such as fuel, food, health care, and housing, declining agricultural productivity, global depletion and contamination of the earth's fresh water resources, oceanic pollution and fisheries depletion, and deforestation resulting from population growth and biofuel production. The impact of biocatastrophe on highly populated urban or coastal regions is enhanced by natural disasters such as hurricanes, earthquakes, floods, ice sheet and glacial melting, and droughts, the intensity of which is often increased by cataclysmic climate change. The most visible of all components of biocatastrophe are the social and economic consequences of the collapse of the viability of the growth-based model of western consumer culture. Unemployment, bankruptcy, falling stock, bond, and real estate values, predatory commodities speculation and equity funds, and a shadow banking network that, in essence, is a form of organized crime (larceny on a global

56

scale,) characterize both the global financial crisis of 2008-? and the eventual collapse of a non-sustainable global military/industrial/consumer society.

X. Health Physics Components of Biocatastrophe

Overview

The health physics impact of anthropogenic ecotoxins is a subject of immense complexity. Modern medical researchers may never have time or funding to begin to unravel and elucidate the relationship between our contamination of the world's biosphere with chemicals we have not even identified yet and their impact on human health. The rapid evolution of numerous antibiotic resistant bacterial strains is an equally complex topic, and of particular difficulty since thousands of varieties of bacteria and other microorganisms have yet to be identified. Even more disturbing is our propensity to dispense growth hormones and other pharmaceuticals and genetically modified organisms without the vaguest idea of the future consequences of these actions. In this context, the global financial crisis based on the careless accumulation of debt is a metaphor for the activities of modern civilization: in a biosphere with finite resources, we have overdrawn our bank accounts of healthy ecosystems, contaminating them with toxic effluents in our frenzied attempt to grow and prosper. A few comments follow on some of the most important health physics issues of this ominous legacy to our children and grandchildren.

Health Impact of Chemical Fallout

The global transport and biomagnification of anthropogenic ecotoxins into and within the myriad food webs of the biosphere has the potential to impact all communities regardless of economic status. The incorporation of ecotoxins into the food chain occurs by bacterial transformation, ecotoxin absorption into, or adsorption onto, organic matter, and the recycling of ecotoxins by the activities of detritivores (scavengers that feed on dead plants and animals or on their waste). Nutrient elements contaminated with ecotoxins are incorporated in living protein by primary producers; ecotoxins are biomagnified as they are transported throughout the biosphere, becoming potent carcinogens, mutagens, teratogens, neurotoxins, and endocrine disruptors. We are just beginning to unravel the secrets of the role microorganisms play in recycling anthropogenic ecotoxins. These ecotoxin exposure pathways culminate in the direct bioaccumulation of environmental contaminants by humans and other animals by ingestion, inhalation, absorption, or cross placental transfer.

Ecotoxin Uptake by Children

The rapid proliferation and global transport of neurotoxins, often in the form of persistent organic pollutants and endocrine disrupting chemicals, has resulted in worldwide contamination of the food chain and the biomagnification of ecotoxins in human blood, breast milk, and tissues. Infants and children are the populations most vulnerable to the effects of neurotoxins, immunological disrupters, carcinogens, mutagens, teratogens, and other ecotoxins as manifested in the rapidly increasing incidence of cancer, endocrine abnormalities, feminization, developmental abnormalities, autism, attention deficit hyperactivity disorder (ADHD), bipolar personality, obesity, and diabetes. Children are the canaries in our coal mines at the dawn of the era of biocatastrophe.

Children as Bioindicators

An important component of ongoing denial of the accelerating contamination of the world's ecosystems by anthropogenic ecotoxins, including persistent organic pollutants (POPs) and other endocrine disrupting chemicals (EDCs), is the scientific community's inability to communicate to the general public their extensive documentation of a broad spectrum of ecotoxins in the blood and tissues of almost all children tested. There has been a failure in media coverage of the links between ecotoxic body burdens of endocrine disrupting chemicals (EDCs), their vast epigenetic impact (changes in gene expression,) their hormone altering potential, and the rapidly accelerating incidence of autism, behavioral and learning disorders, ADHD, and other endocrine-related problems in children. The continuing assertion by many in the medical community that we don't understand the etiology of many of the emerging disorders recently observed in both children and adults is consistent with the lag time between effluent discharge and the minimal efforts at documenting their health physics impact. Our lack of knowledge of the consequences of our actions, along with economic and political considerations, becomes an excuse for the continuation of the current practices of the manufacture and distribution of new variants of ecotoxic chemicals without first studying their health effects. One of the more blatant examples of this risky practice in the United States is the proliferation of the use of brominated fire retardants (the PBDE family of ecotoxins), which were long ago banned in Europe.

Children and Thermisol

The rapid acceleration in learning and behavioral disorders in children has been linked by some autism activists to the methylmercury in thermisol, a preservative contained in vaccines. The use of thermisol has been discontinued in vaccines in the US, but the incidence of autism continues to rise. No scientifically verifiable linkage between the use of thermisol and autism has been established, yet methylmercury, which has many point sources, including the burning of coal, is now appearing in many processed food products. Of particular interest is its appearance as a microcontaminant in high fructose corn syrup (HFCS); it is also ubiquitous in the dietary intake of mothers and children, including in breast milk, sea foods, and numerous other pathways to consumption by children. As a potent neurotoxin and one of many EDCs widely documented in food webs, the symptoms of mercury toxicity share a striking similarity with the symptoms of autism (see *Appendix P* in *Volume 3*). Methylmercury is only one neurotoxin among many ecotoxins to which children are exposed by their dietary intake. The larger issue of the intake of and exposure to ecotoxins by children is the subject of continuing denial by a majority of Americans – parents, educators, and the medical community – who are unable to acknowledge, often for political, religious, or economic reasons, the link between the declining health of children and the increasing contamination of the ecosystems in which they live. The accelerating expense of healthy whole foods in a world of highly processed foods with questionable additives, a broad range of environmental microcontaminants, and vast income disparities among consumers highlights the challenges faced by most families to provide their children with a healthy diet. The fact that maternal cord blood and breast milk are universally contaminated with anthropogenic ecotoxins is an unfolding and tragic complication with respect to the challenging task of providing healthy food to children (see the last appendix in *Volume 3*).

Antibiotic Resistant Bacteria (ABRB) and Newly Emerging Pathogens

The worldwide overuse of antibiotics by well-intentioned health care professionals has resulted in the proliferation and global transport of antibiotic resistant organisms such as MRSA, *Clostridium difficile*, vancomycin resistant *enterococcus* (VRE), antibiotic resistant TB, Lyme disease, and other developing and mutating bacteria. Overuse of antibacterial cleaning agents, including the increasing use of nanosilver, in both domestic and medical environments is also a source of increased bacterial resistance, reducing the future effectiveness of antibiotics in increasingly vulnerable populations of urban and institutional environments. The elderly, pregnant mothers, newborns, users of mass transportation systems, and patrons of sporting and entertainment events are at most risk as outbreaks of ABRBs, such as MRSA and *C. difficile*, grow from local occurrences to regional epidemics and global pandemics.

Newly Emerging Pathogens

Increasing world population, combined with global trade, modern transportation systems, and consumer product exchange networks, promote rapid hemispheric transport of previously isolated pathogens. Avian flu, SARS, norovirus, Ebola, Marburg virus, bovine encephalitis, and other microorganisms with growing animal to human transfer capabilities (zoonosis), as well as unidentified new mutations, such as the recent outbreak of H1N1, and reemerging diseases, have the potential to affect population groups at all socio-economic levels. The decimation of a majority of cattle in the United Kingdom in the 1980s as the result of the spread of Mad-Cow Disease is one

60

example of the emergence of a pathogen with vast economic and social consequences. The huge assembly-line stockyards (CAFOs) of the American meatpacking industry and their growth hormone- and pesticide-laced inhabitants, already frequently infected with salmonella, provide a fertile environment for the spread of a variety of reemerging or newly-emerging pathogens in pathways to human consumption. In a worst case scenario, much of America's meat supply could be cut-off within a few months by the unexpected emergence of one or more pathogens that would discover the massive and vulnerable Midwestern stockyards as an ideal environment for proliferation.

Variants of Newly Emerging Pathogens

The recent emergence of Panton-Valentine leukocidin as a component of MRSA results in a bone marrow-consuming infection in children. This newly emerging pathogen is an example of the evolution of a pathogen that has the genetic capability to merge with an already well-established disease, which itself is undergoing genetic mutations in response to the attempts of the medical community to control or eradicate its presence, especially in hospitals where it has been present for decades. The gradual expansion of the frequency of appearance and increased potency of MRSA in American and world hospitals and its emergence as a community pathogen is complicated by its growing resistance to antibiotic treatments and the potential for rare synergistic variants to emerge, as with the leukocidin-MRSA variant.

TB as an Example of a Reemerging Pathogen

Formerly dormant or previously controlled pathogens such as TB (tuberculosis) and smallpox are particularly prone to reemergence as a public health threat in a world characterized by a global consumer culture with efficient worldwide transportation networks. The recent emergence of MDR-TB (multi-drug-resistant TB) has been followed by the appearance of XDR-TB, an even more drug-resistant form of TB. The potential for the worldwide spread of both multi-drug-resistant forms of TB has yet to receive significant mass media publicity despite its global proliferation, which has recently reached a half million cases a year. The recent documentation of the spread of HIV-TB coinfection adds one more chapter to the growing history of TB as a reemerging threat to public health. Tracking the worldwide spread of new forms of antibiotic resistant TB from current loci of epidemics in Eastern Europe and Africa is already one of the top priorities of the Centers for Disease Control (http://www.cdc.gov/tb/default.htm).

Other Reemerging Pathogens

Previously isolated infectious biological agents, including viruses (Marburg, Ebola), newly-resistant bacteria (antibiotic resistant TB, SARS, Acinetobacter baumannii), reemerging diseases (smallpox), and age-old threats to world health, such as malaria, can be easily and rapidly transported by anthropogenic activities in a biosphere that is getting smaller in proportion to rising human population. The regional spread of pathogens, such as cholera, is limited to wastewater/drinking water pathways; long distance and worldwide spread of local epidemics is unlikely. More ominous is the potential for the rapid spread of mutated forms of viruses, such as Avian flu, which is closely related to conventional flu viruses. The growing number of viruses that have the potential to cause a worldwide pandemic, combined with the ease of transport in a globally-connected world trading culture and the rapidly increasing size of vulnerable urban environments provide the scenario for a potential human health catastrophe. The probable future development of weaponized viruses in a world of social unrest further complicates this scenario. Only the rapidly advancing understanding of the genomic constituents of those microbial threats and the consequential development of effective vaccines (But for how long a time period will a particular vaccine be effective?) may offset the proliferation of their diversity.

Malaria as a Paradigm

The recent emergence in Cambodia of drug resistance in a parasite that is the vector for the deadliest form of malaria is one example of the multiplicity of the health threats that are a growing component of globalization. The Bush administration invested 1.2 billion dollars in its Malaria Initiative in an attempt to reduce the death rate of 2,000 African children per day (Fuller 2009, D1). The most recently developed anti-malaria drug, artemisinin, extracted from an ancient Chinese herbal medicine, has been the most effective of modern anti-malarial drugs, but has recently shown a loss of potency in combating newly emerging strains of malaria-causing parasites in Cambodia (Noedl 2008). The potential spread of this artemisinin-resistant strain outside of the Cambodia/Thailand area constitutes a major threat to world health in all tropical 3rd world countries as well as in more advanced western and Asian communities. It is one example of emerging health threats in a world economy characterized by the increasing ease of pathogen transportation, delayed development of new treatments, and the declining availability of world financial resources to combat emerging health threats.

Genetically Modified Organisms

The phenomenon of biocatastrophe is further complicated by the consequences of the inadvertent production and distribution of mutagenic and teratogenic chemicals, which cause the proliferation of pesticide- and herbicide-resistant primary producers (plants and algae), and the evolution of genetically altered organisms. In the lower levels of the food chain, the introduction, proliferation, and circulation of ecotoxins result in the emergence of genetically modified heterotrophic microorganisms (organism consuming

62

organisms), which are then subject to global transport mechanisms. These new forms of bacteria, fungi, and other microorganisms have the potential to permanently alter the well established, natural balance of microorganisms in stable ecosystems, as well as facilitate unexpected epidemics in urban environments inhabited by humans, especially as they become established in vulnerable but receptive environments, such as hospitals, nursing homes, penal institutions, and military facilities. The spread of a variety of staphylococcus and viral infections from Kuwait and Iraq to the United States by U.S. servicemen occurred during the First Gulf War period. The etiology of many of these infections may be connected to the use of highly toxic chemicals, including contaminants in the anthrax vaccines given to veterans. The experience of Gulf War veterans alerts us to the complexity of the relationship between anthropogenic chemicals and human health. The contaminated anthrax vaccines given to Gulf War soldiers is one specific example of the unexpected health effects of the well intentioned efforts of the military medical establishment to protect our troops in what is now regionalized global warfare, an unfortunate event recapitulated in domestic institutions by recent outbreaks of MRSA and *C. difficile* infections.

For-profit Genetic Engineering

For-profit genetic engineering, which facilitated the first green revolution and the consequential explosion in world population between 1960 and 2005 also has the potential to result in long term decline in the genetic diversity of the world's horticultural resources and the food production capacity of sustainable, self-sufficient economies. Genetically modified seed crop varieties now dominate most seed distribution networks. Genetically engineered corn, one variety of which tastes so succulent, in contrast to the more widely produced unpalatable herbicide resistant corn, produces higher profits for farmers and agribusiness in a short term cost-benefit analysis. The future costs of the loss of the genetic diversity of crops, such as corn, cannot yet be calculated. The recent accidental distribution of genetically engineered corn, called "event 32," is one example of the inadvertent spread of a genetically altered species that may have a future impact that cannot be now evaluated. Genetic drift of newly-designed food products has the potential to alter the worldwide gene pool. The development of industrial agriculture and the decline in local sustainable agriculture results in the inadvertent spread of invasive and/or altered species, a decline in the diversity and availability of long-established crop species, and the potential for the rapid decline in the productivity of large areas of the world's food-producing croplands by the global transport of genetically altered autotrophs. The probable long term external costs of genetically engineered food crops have the potential to overwhelm the more well known benefits of the development of fungus and virus resistant crops, vitamin-enhanced crops, and drought-resistant species, all of which have the potential to benefit struggling indigenous nonindustrial agricultural communities.

Pharmaceutical Ecotoxins

Among the most notable of all pharmaceutical ecotoxins are the antibiotic resistant bacteria (ABRB) now the ubiquitous product of the misapplication of antibiotics for both human health needs and for treatment of infections in animals. Obviously of significant initial efficacy, the use of antibiotics has and will continue to save millions of lives. Unfortunately, antibiotics are frequently misapplied to treat viral infections, or over-prescribed, misused, or excessively ingested in humans and livestock, creating numerous forms of antibiotic resistant bacteria. The ever increasing use of antibiotics to treat mastitis and other infections in cattle, chicken, and other livestock further exacerbate the proliferation of ABRBs. The same observations may be made about a wide variety of initially useful or hopefully useful pharmaceuticals. Once disposed of in wastewater treatment systems, antidepressant medications, anti-cancer drugs, erectile dysfunction medications, sex hormones, birth control pills, and a variety of other medications are transferred to public water supplies and become potentially biologically significant ecotoxins. Pharmaceutical wastes are quickly transported by the global atmospheric water cycle and now contaminate most streams, rivers, surface waters, and soils in all but the most isolated locations (see *Appendix R* in *Volume 3*). The long term impact of the use of growth hormones, such as rGBH, originally administered to cattle and chickens to increase body weight and, thus, profits, and now a ubiquitous contaminant of regional and local water supplies, increases the production of the hormone IGF-1, a naturally occurring hormone, which in elevated concentrations has been linked to the development of cancer (especially breast and prostate cancer,) loss of bone mass, and increase in risk of cardiovascular diseases. Other health physics impacts are as yet undocumented, but the worldwide use of growth hormones, and in America the continued use of rGBH, constitute the tip of the iceberg of the proliferation of economically, socially, and medically useful pharmaceuticals, which may have delayed biological significance as anthropogenic ecotoxins.

The Synergism of Multiple Pathogens and Ecotoxins

The phenomena of newly emerging and antibiotic resistant bacteria (ABRB) may be exacerbated by increasing environmental inventories of chemical and pharmaceutical ecotoxins that may also have a synergistic impact on their growth and proliferation. Ecotoxin synergism may exacerbate the vulnerability of genetically altered food crops to new pathogens and the impact of the overuse of antibiotic medications. Ecotoxin synergism may also enhance the adverse impact of the use of antibiotic and growth hormone additives in animal feed, use of antibacterial cleansers, and the deliberate and

64

accidental proliferation of synthetic hormones in consumer products and food webs. The proliferation and dispersion, especially through contaminated surface and public water supplies, of a variety of antibiotic resistant bacteria, newly mutated viruses, genetically modified organisms, and toxic pharmaceuticals have the cumulative effect of disrupting the cybernetic networks (control systems) of natural and human ecosystems.

Health Impact of Genetic Engineering

New forms of genetically modified organisms (GMOs) designed for short term economic gain in industrial agriculture, for treatment of human health conditions, or to combat newly emerging pathogens have the potential to facilitate antibiotic resistant bacteria (ABRB) interactions and the creation of variant forms of GMOs and other pathogens. Newly bioengineered GMOs have the potential to cause unanticipated side effects or otherwise become unexpected public health threats. In this context, increasingly potent antibiotic resistant bacteria such as *C. difficile*, after contact with and exposure to a genetically altered organism carrying other types of ABRBs, could mutate into an entirely new form of ABRB. The potential variations in the etiology of new forms of ABRB as a result of their contact with new variants of genetically modified organisms and foods are nearly endless. The ingestion of genetically modified foods has the potential to alter human DNA, increasing the risk of the evolution of new unforeseen threats to future generations.

The Future of Children

Environmental ecotoxins and their link to autism and learning disorders cannot be separated from the changing lifestyles characteristic of global consumer society, including the increasing consumption of mass-produced, processed foods often made with high fructose corn syrup and hydrogenated vegetable oil. Increasing rates of obesity and diabetes are part of the declining health of children in the age of rapidly advancing health care strategies and technologies. In view of the future impact of the rapidly growing cost and possible unavailability of healthy, whole foods and potable water, and the continuing increase in the variety and intensity of ecotoxin contaminant pulses, the health status of the children of the future is uncertain. For children of low income parents the future is grim. The rapid spread of the shortage of potable water, increasing food stress and malnutrition, the frequent reoccurrence of cholera epidemics, the threat of emerging drug-resistant strains of malaria, and the emergence of HIV/TB co-infection on the African continent are among the most ominous contemporary health physics issues at the dawn of the Age of Biocatastrophe. Children are the first victims of these threats. Most unfortunate of all is that there may be no solution to these and many other challenges to the future health of the majority of the world's children, especially as the world financial crisis undermines the availability of funds for public health initiatives and management of health care crises.

Declining Life Expectancies

Chernobyl, a regional biocatastrophe, was one probable cause of the sudden drop in the life expectancy of residents of Belarus and the Ukraine. In the US, for the first time in its history, there has been a drop in the life expectancy of low income women in certain Appalachian States and other selected regions. The onset of biocatastrophe and its progression will be characterized by the declining life expectancy of an increasingly larger proportion of the world's population. The possibility of a sudden decline in world population, the ultimate consequence of the population bomb, has long been predicted. What is now emerging is an awareness of the link between global economic insecurity, declining ecosystem biodiversity and productivity, and the increasing risk of sudden population decline. If and when the world economy collapses, population die-offs will soon follow.

Public Health Assessments

The study of the public health significance and etiology of the adverse health effects of exposure to multiple ecotoxins in children and adults is dependent on the accurate measurement of their bioaccumulation, biomagnification, and synergistic interaction in abiotic and biotic media. The urgent need to measure increasingly strong contaminant signals in all media is undermined by the political and religious denial of their presence and the economic reality of declining public resources in the context of increasing public needs. A fundamental conflict of interest exists between the economic, political, and religious values of the American free enterprise system and the global consumer society it engendered, and the need to assess and mitigate the public health impact of its activities. This ongoing crisis forms the context for the ongoing denial of an emerging public health crisis. The rapidly rising health care costs of this accelerating public health care crisis are only one component of the rapidly unraveling social and economic milieu of American and global consumer society. The one positive note in the sorry tale of the failure to control the rapid spread of anthropogenic ecotoxins in animal and human populations is the robust and accelerating documenting efforts of the US Centers for Disease Control (CDC), which has recently issued its fourth survey of ecotoxins in humans. When the Tea Party nitwits insist on dismantling the EPA as part of their anti-government agenda, would they also do away with the CDC and its growing documentation of the tragedy of our Round-World Commons?

Health Care Disparities and the Unexpected World Financial Crisis

Our sketch of the phenomenon of biocatastrophe has been written in the context of the last two centuries of environmental impact of human activities, as well as the projection of probable scenarios during the coming decades, if not centuries. The full impact of a catastrophic rise in sea level may not be felt for a century or longer. In the United States, the unexpected collapse of the viability of our commercial financial institutions has occurred in just one year. Gradually rising personal, corporate, and state debt, which began in the Reagan era, has culminated in current account deficits on a state and national level. Single year budget shortfalls in most states now average approximately $1,000 per person. The national balance of payments deficit for 2009 may average $4,000 per US citizen; for 2010 it will approach $5,000 per citizen. These are the shortfalls for one fiscal year. In the current contest for full access to the benefits of modern medicine between the super rich, the techno-elite, and the insured members of the middle class, versus the uninsured, most children, and the poor, a satisfactory resolution of the glaring disparities in health care availability has not yet been resolved despite the recent modest success of the Obama administration in passing health care reform legislation. The rapidly deepening economic crisis of most state governments is recapitulated in growing family indebtedness, not to mention a ballooning federal deficit. The recent passage of health care legislation only partially mitigates the glaring health care availability disparities that characterize America's unique "every person for themselves" antigovernment political milieu. In the context of the sudden appearance of an unexpected world financial crisis, now centered primarily in the developed nations of our western market economy, how can adequate health care for all US citizens be financed without increasing our collective indebtedness? In view of the widespread public antagonism towards increasing Medicare taxes or otherwise reforming our dysfunctional tax codes, what are the alternatives that could eliminate our glaring health care availability disparities? Better health care for 3rd world populations? Don't even ask.

XI.   The Politics of Biocatastrophe

The Roots of Biocatastrophe

The roots of biocatastrophe lie in the gradual evolution of western market economies beginning with the emergence of the Italian city-states in the 14th and 15th centuries and the robust trading economies of the expanding western European empires of the 16th and 17th centuries. The world exploration that began in 1492 was initially a search for a shorter route to the precious metals, silks, and spices of India and the Far East. The organization of the south and north Virginia plantations (1606) occurred at the same time as the formation of the East India Company. No shortcut to the riches of the Asian continent was ever found; the furs, fish, and forest products of North America were soon discovered to be a most valuable group of commodities for the rapidly growing market economies of European empires. Only three decades after John Smith discovered that codfish were the biotic equivalent of gold (1614), recently arrived New England shipwrights began building East India merchantmen in Salem and Boston. America's unique free enterprise system, based in part on the unlimited horizons of western expansion, was off to a running start. One of the first acts of English fur trappers in New England was to completely harvest all the beaver populations without regard to maintaining a population of breeding females, a portent for the inalienable right of members of an incipient free enterprise system to exploit the natural resources of the new-found-lands without regard to the future consequences of their actions.

The Roots of the American Free Enterprise System

The evolution of the uniquely American consumer culture, fully developed by the early 1960s, was characterized by the hijacking of the ideals of Jeffersonian democracy and the US Constitution and their amalgamation with the primitivism of evangelical Christian fundamentalism. A key component of the birth and growth of a sectarian philosophy of a uniquely American free enterprise system was the promise of exemption for all evangelical true believers from the consequences of aberrant behavior, i.e. the ideal of Christological innocence. The quotation from Genesis cited at the beginning of this volume provides the religious rationalization for what is an entrenched attitude of the American "free" enterprise system. The combination of the religious, political, and social ideologies underlying a specifically American philosophy of free market economics provides the basis for the emergence of Reaganomics after 1981, an anomalous late 20th century version of completely unregulated private sector business activities as one of the inalienable rights guaranteed by the US Constitution. This sectarian ideology, enthusiastically embraced by many Christians as well as non-Christians, includes the inalienable right to the unlimited accumulation of property,

68

consumer products, wealth, and other assets. It also includes the ideology of the inalienable right to possess firearms of every caliber and description, and the cult of power as expressed by the ownership of wealth, ever-growing stock market portfolios, real estate assets, vehicles, and consumer products.

A Uniquely American Ethos

The ideal of the unfettered right to contaminate the biosphere with ecotoxins produced by a profit driven free enterprise economy is a natural outgrowth of the evolution of this uniquely American ideology. Its recent manifestations include the emergence of a predatory and highly profitable shadow banking network, characterized by sophisticated financial assets manipulation by now infamous Wall Street "quants" (Patterson 2010), and a narcissistic cult of conspicuous consumption that has its roots in the Bronze Age empires of Mesopotamia and Egypt. Wannabe Cleopatra's infest America's labyrinth wasteland of cable channel nightmares, among the most telling symptoms of a consumer culture with terminal acquisition syndrome disorder. This ethos is best summarized as the American Environmental Fascist Free Enterprise System (AEFFES). One might also use the term Christian-Environmental Fascists in the context of the recent evolution of a vocal far right fringe element in mass media represented by Rush Limbaugh (a.k.a. Mush Rambo) and other pundits (Hannity, O'Reilly, Levin, Ingraham, Savage, Beck, Coulter, and many others). This designation is unfair to most Christians, the quiet urban, suburban, and rural family centered backbone of once sustainable American culture, church-goers of every class and community. America's vocal banana republic media pundits and commentators represent only a small percentage of American citizens and do not represent the majority of Christians, Quakers, Jews, and free-thinkers whose values are the antithesis of the American Environmental Fascist Free Enterprise minority. The values of this vocal minority have been clearly articulated by the growing, if diverse, Tea Party movement as well as by the recent (February 2010) vocal Conservative Political Action Conference (CPAC convention). They also function, in essence, as effective spokespersons for America's gun-toting white supremacists, Ayan Rand advocates, and, though unintentionally, for the predatory community of Wall Street financial engineers and their collateralized debt obligation (CDO) and credit default swap (CDS) scams.

A Definition of Fascism

The *Concise Columbia Encyclopedia* (3rd edition) published by the Columbia University Press (1994), contains the following definition of fascism: "A philosophy of government that glorifies the nation-state at the expense of the individual. Major concepts of fascism include opposition to democratic and socialist movements, racist ideologies, such as anti-Semitism, aggressive military policies, and belief in an authoritarian leader who embodies the ideals of the nation. Fascism generally gains support by promising social justice to discontented elements of the working and middle classes, and social order to powerful financial interests. While retaining class divisions and usually protecting capitalist and land-owning interests, the fascist state exercises control at all levels of individual and economic activity, employing special police forces to instill fear." Vice President Henry Wallace made the following comments about "American Fascism" in a 1944 article in *The New York Times*.

> If we define an American fascist as one who in case of conflict puts money and power ahead of human beings, then there are undoubtedly several million fascists in the United States. There are probably several hundred thousand if we narrow the definition to include only those who in their search for money and power are ruthless and deceitful. . . . They are patriotic in time of war because it is to their interest to be so, but in time of peace they follow power and the dollar wherever they may lead. (Wallace 1944, SM7)

Environmental Fascism and Free Enterprise

With small adjustments in terminology, there is a close relationship between the fascism of Nazi Germany, Mussolini's Italy, or any run-of-the-mill banana republic, and the environmental and free enterprise fascists who play such an important role in the genesis of biocatastrophe and the suddenly unfolding global financial crisis. If we substitute corporate state (read: consumer state) for nation-state, the prime function of the individual is to be a consumer in the tradition of Wal-Mart Woe-Man (WMWM) in the context of a totally unregulated free enterprise system. Major concepts of environmental and free enterprise fascism include opposition to any limitations on the inalienable right to consume infinite quantities of natural resources, to dispose of immeasurable quantities of ecotoxins into the biosphere, and to accumulate limitless quantities of consumer products, real estate, stocks, bonds, or money. Racist ideologies continue to characterize environmental and free enterprise fascism in the form of a not so subtle continuation of traditional attitudes of racism, disdain for third world non-participants in the global consumer culture, and opposition to sustainable economies and environmental awareness. Aggressive military policies continue their universal appeal, at least with a large minority of American citizens. The authoritarian leaders who embody the ideals of this uniquely American ethos are the talk show hosts, pundits, commentators, and highly visible evangelical ministers who effectively use mass media to proselytize; political leaders, such as Ronald Reagan, who effectively articulated the values of the new ideology, and sports idols, movie stars, and pop culture heroes who glorify the ideal of the conspicuous consumption of nonessential luxury goods. The recent election of Barack Obama provides hope for the mitigation of the virulence of this ethos in the context of a protest by a small majority of Americans against the continuation of its influence. The promise of the possibility of the unlimited future consumption of consumer products is a psychotherapeutic substitute for the promotion of social justice, enhances America's growing reputation as the homeland of mass culture and functional illiteracy, and supports the need for financial and land owning interests to maintain social order. The global consumer culture and its obsessive need for ever-increasing growth and consumer product consumption controls all levels of individual and economic activity. Special police forces in the form of a wide variety of federal, state, local, and private security apparatus proliferate in our post-911 world of jihad, environmental fascism, and growing social unrest, ideally insuring that terrorists and criminal factions do not interfere with the efficient functioning of what was, until Nov. 6th, 2009, appropriately called the American Environmental Fascist Free Enterprise System. Will the Obama administration represent a historic turning point where the values and destructive impact of non-sustainable consumer society, revealed by the epiphany of America's September 2008 financial collapse, will be the subject of public discourse and debate?

The American Environmental Fascist Free Enterprise System

Only a tiny percentage of Christians participate as active ideologues in the AEFFES. Most enthusiasts don't go to church on Sunday. A surprising number of foreigners are its often unwitting supporters, including non-Christians such as Indian businessmen, Japanese industrialists, Chinese traders, Jewish Ponzi scheme perpetrators, agnostic Swiss bankers, German entrepreneurs, Icelandic money managers, and English investors. The fundamental beliefs of hard core participants in the AEFFES are the inalienable right to accumulate unlimited wealth and property in a biosphere of limited resources, devotion to the cult of money and power, and, in the case of a large majority of American participants, the worship of the ideology of the unrestricted ownership of firearms and a concomitant, not always clandestine, reverence of violence. When Bernie made off with billions from Jewish philanthropies, European banks, the palisaded techno-elite at the Palm Beach Country Club, and mom and dad's mutual funds, his convivial demeanor and long years of highly visible Wall Street experience helped make him one of the more successful participants in Wall Street's predatory free enterprise shadow banking network. This milieu, based on the marketing of debt as an interest-paying security, was born as a child of Reaganomics in the post-Kennedy era of the decline of American industrial and political hegemony. The gun toting enthusiasts of the AEFFES and its growing post 9/11 security apparatus are its most visible icons. Less visible but essential to its formerly invincible economic hegemony are the creative financial engineers and their culture of repackaged debts marketed as AAA collateralized debt obligations (CDOs) and then insured by greedy financial institutions as highly remunerative credit default swaps (CDSs). This financial market cancer, which David Gergin appropriately called America's "casino culture" on CNN, possibly life-threatening to the viability of the global market economy, allowed a select few to make-off with the assets of the hapless many. One challenge for the Obama administration is to help mitigate the mass functional illiteracy that is, in part, the wellspring of the power of this component of American society.

The Death of Camelot

The assassination of President John F. Kennedy on November 22, 1963, at the height of the Age of Chemical Fallout, marked a turning point in American history. Vietnam, the Chicago riots, the deaths of Martin Luther King and Robert Kennedy, Watergate, the Iran hostage crisis, all followed. The era of Reaganomics (1981-2007) marks the culmination of the successful hijacking of the American free enterprise market economy, whose roots lie, in part, in New England's cod fishing, West Indies trading, shipsmithing, and shipbuilding industries. The classic period of American toolmaking (1827-1930), when the family business flourished in the growing shadows of omnipresent robber barons, tycoons, and unregulated, untaxed monopolies that foreshadowed the ideals of modern "live free or die" Republicanism, was a prelude to the final domination of the ethos of American culture and its economy by a consumer culture that gradually mushroomed over our physical and psychological landscapes after 1945. The optimism of the Truman/Eisenhower/Kennedy era ended in the sectarian ideology of the Bay of Pigs and the Cold War necessity of the supremacy of state security apparatus. The end of the Cold War marked the beginning of the hegemony of a rapidly expanding American Environmental Fascist Free Enterprise System (AEFFES) based on narcissistic consumption, growing income disparities, and the illusion that the galloping indebtedness of American consumer society could be paid for by an ever expanding global consumer culture utilizing the dwindling assets of an ever-shrinking biosphere. The terrorist attacks of 9/11/01 marked the beginning of the unraveling of the edifice of Reaganomics; the death knell of the lucrative shadow banking scams of the AEFFES was sounded by the collapse of Lehman Brothers and the world financial crisis of 2008-?. A late middle age global consumer society may flourish again, but only for a few decades and always based on the ephemeral illusion that we, not bacteria, are the species in control of Gaia.

Anger and the Evolution of Tea Party Politics

The election of the Obama administration in 2008 signaled the temporary end to the era of American theocracy, at least with respect to the influence of a militant evangelicalism, self defeating political unilateralism, and the mindless accumulation of personal, corporate, and banking debt as described in a book of the same title by Kevin Phillips (2006). The Bush era was also characterized by the increasing power of

72

national and multi-national corporations, the growing influence and unfortunate impact of an out of control shadow banking network, and the ill-fated war in Iraq, which provided a huge impetus to the worldwide growth of Jihadist militancy. The growing American "disenlightenment" that Phillips describes (2006) as characterizing American politics at the end of the ear of Reaganomics culminated in the floundering directionless of the Republican Party that resulted in the anomalous election of the Obama administration. The spectacle of growing unemployment, economic dislocations, ballooning debt, and a well paid and unregulated Wall Street has resulted in the rapid evolution of "Tea Party politics," an uninformed, reactionary, anti-progressive coalition, which, in spirit, represents the broad diversity of all the conservative political elements in American society. Sparked by America's suddenly visible economic crisis and motivated by long standing antagonism to government regulation of any type, the Tea Party movement symbolizes the contradictions between the necessity of funding warfare in Vietnam, Iraq, Afghanistan, and elsewhere; the legacy costs of Social Security, Medicaid, and Medicare; and the desire to maintain minimal federal tax rates. That these frustrations about the sudden intrusion of "big government" into the lives of ordinary citizens have suddenly resurfaced after the election of a charismatic, progressive, democratic president of color indicates the strong racial bias of much of the Tea Party movement. Intrusive government was not an issue with the conservative movement during the Bush administration's ill-advised invasion of Iraq or its huge budget deficits. One word serves to describe the emotional state of millions of Americans who would be much happier if they were not economically marginalized by the unfortunate interface of the sophisticated world of information technology and the predatory world of corporate America: Anger. The Obama administration, as well as the vacillating Democratic party are faced with the near impossibility of implementing the progressive legislative reforms essential to the future viability of American society: comprehensive health care legislation, regulation of an out of control banking system, greenhouse gas reductions, and many other progressive reforms. Tea Party fascists from both parties effectively join with traditional conservatives, radical evangelicals, reactionary media pundits, and other sectarian and partisan elements in American society to vote NO on any and all proposals that might ameliorate America's fall from the status of a world power to that of a colony of developing countries now best known for their adroit adaptation of America's creative information age innovations.

2010 Update: The Ominous Unraveling of Civil Discourse

Much of the optimism that the Obama administration represents a new age for American democracy is now fading. Special interests have succeeded in sabotaging what would have been important components (e.g. a single payer system) of the recently passed health care reform legislation. The intense partisanship of a dysfunctional Congress and the economic dislocations of the ongoing recession are mirrored in the rapid rise of crypto-fascist anti-environmental and racist anti-Obama Tea Party enthusiasts. FOX News has become by far the most popular and influential electronic news media outlet; Glenn Beck has replaced "Mush Rambo" as the most visible spokesman for a growing anti-government movement that now dominates the public non-debate of the critical issues of the time. The Copenhagen Climate Conference collapsed in dystopia; anti-green Al Gore bashing literature makes frequent appearances on the shelves of large bookstores. The lack of civility that characterizes Congressional, electronic media, and political discourse is not limited to the revival of America's most reactionary conservative elements; the language in Huffington's *Pig's at the Trough* (2004) sets the tone for most commentary on the rise of America's emboldened Tea Party fascists. While a few voices remain to provide an alternative to FOX News (*The New York Times*, National Public Broadcasting, Alternative Radio, Democracy Now), incisive analysis of the rapid decline of the non-sustainable elements of America's narcissistic consumer culture is almost nonexistent. The recent Supreme Court elimination of restrictions on corporate and special interest election advertising further undercuts the possibility of informed debate via the medium of mass media outlets.

XII. The Economics of Biocatastrophe

World Food Supplies

The early stages of biocatastrophe are and will be characterized by increasing costs for basic food stocks and growing interruptions in food supply availability. First impacting the invisible underclass of third world communities, the cost of basic food supplies has begun and will continue to restrict the dietary options of lower and middle income families on all continents. Increasing incidents of food rioting, theft, food hoarding, and food supply interruptions characterize the gradually declining affordability and availability of basic food stocks. In the context of the suddenly unfolding world financial collapse of 2008/? with its rapidly rising unemployment and spreading family bankruptcies, many of these developments are already afflicting communities in America and western Europe. Oceanic fisheries depletion, climate change-induced drought, soil depletion, salinization, increasing energy-related transportation costs, and the proliferation of ecotoxin-laced mass produced "fast foods" utilizing high fructose corn syrup (HFCS) and hydrogenated oils will greatly exacerbate future global food shortages by dramatically reducing the quality of food available to most world citizens who are now or will become dependent on global food production networks. Unexpected disruptions in the global market economy and its financial institutions will further enhance the shortages that result from the collapse of human and natural ecosystems that once supplied whole foods in sustainable economies.

Food Insecurity

Press reports indicate the emergence of a consensus that at least one billion people are now at risk of starvation. The media term for this phenomenon is "food insecurity," now besetting 20% or more of the U.S. population. The world financial crisis marks the beginning of a new age of media attention to the plight of hundreds of millions of malnourished Third World people who have long been the victims of a world food crisis, as well as the equally large population impacted by growing regional warfare and the consequences of living on the margins of a global economy where unemployment is a way of life. Increasing attention is now been given to the challenges faced by the many communities that cannot afford to participate in the global economy. The sustainable lifestyles of these communities have been undermined or destroyed by global warfare, ecosystem destruction, resource depletion, rising commodities prices, financial insecurity, the intrusion of corporate entities into local economies, and the increasing dependence on remote sources of basic commodities. Natural disasters such as the Haiti earthquake graphically illustrate the vulnerability of Third World communities to infrastructure-destroying events. There is now a growing public awareness that something is out of kilter in the daily routine of our everyday lives. This growing awareness was recently expressed in Thomas L. Friedman's article "The Inflection is Near" (2009), which contrasts sharply with the previous myopia of his "world is flat/information technology benefits all" thesis (2006), though the word inflection doesn't elucidate the real issue, the ongoing collapse of American-sponsored non-sustainable global consumer society and its intricate pyramid of national and world indebtedness. The economic and psychological consequences of the disintegration of a predatory and unstable global market economy have a worldwide impact on the vast majority of citizens who live outside of the shrinking periphery of its beneficence.

Agricultural Productivity

Human populations need a minimum of 200 acres of productive farmland for every acre of urban environment. Rapidly expanding human populations and their increasing participation in a global consumer society are reducing the availability and productivity of croplands, grazing areas, and other food producing environments. Urbanization, deforestation, soil degradation and erosion, salinization, the world water crisis, population growth, and the intrusion of ecotoxic POPs all contribute to declining agricultural productivity. The looming shortage of phosphate fertilizers (circa 2100) and the increasing cost of essential energy resources will be additional factors limiting ecosystem productivity. Near maximum productivity of existing croplands has already been achieved by the highly mechanized fossil-fuel-dependent "green revolution," its super-productive seed varieties, and the widespread use of potent nitrogen-based fertilizers, ecotoxic pesticides and herbicides, and its increasing reliance on growth hormones, antibiotics, and genetically modified crops. Newly emerging pesticide-resistant pests will be a further limitation on agricultural productivity. Only the growing world awareness of the viability of a second sustainable "organic green revolution" provides hope that the menace of industrial agriculture can be challenged by local and bioregional networks of sustainable community agricultural enterprises – a fundamental manifestation of a truly free enterprise system.

Accelerating Commodities Prices I

The early stages of biocatastrophe are characterized by increasing prices for basic commodities such as cooking oil, milk, flour, vegetables, and animal products. Increasing costs for motor vehicle fuels, home heating oil, natural gas, and other energy sources exacerbate the general increase in costs for all essential products. Future decreases in the capacity of worldwide food production systems caused by industrialization, urbanization, deforestation, water shortages, the destruction of oceanic fisheries, and the displacement of food production by biofuel production will be among the primary causes of accelerating commodities prices. That the incomes of many families, communities, and states are already insufficient to pay incurred bills provides an unnerving preview of the future economic consequences of the collapse of an unsustainable global economy as commodity prices rise and the relative income of most world citizen's fall.

Economic Impacts on Third World Populations

The poorest third of the world's population, few of whom have average incomes greater than $2 per day, is already affected by rapidly increasing fuel, fertilizer, and food costs. The world price of cooking oil has recently doubled in one year due in part to the disastrous impact of biofuels production. Fertilizer and fuel cost increases greatly exacerbate increasing food prices, eventually impacting all but the palisaded elite. The sudden decline in gasoline and oil prices as a result of the 2008-? financial crisis and the collapse of the speculative bubble that accompanied these declines does not alter the inherent limitations of providing food and fresh water for a needy world population in the context of rapidly diminishing fossil fuel supplies. As the peak in the productivity of industrial agriculture is reached, possibly by 2010 or 2015, and as world water stress increases, the timeframe of the evolution of biocatastrophe may be accelerated. In the long term, the coming of the mature stage of biocatastrophe may be measured in decades, rather than in half centuries. The impact of the increasing price and unavailability of basic commodities on invisible third world populations is already vastly underreported by American media. Aversion to a more vigorous public debate about income and health care disparities has long been a component of the overall denial of the collapse of the sustainability of an increasingly predatory global military/industrial/consumer society. Perhaps America can join Europeans in a more open debate about the economic, social, and political consequences in third world countries of the unfolding specter of the decreasing varieties of increasingly expensive foods that are no longer locally grown in sustainable agriculture communities.

World Food Crisis

The effect of cost increases for basic commodities, such as food, fertilizers, and fuel, on middle class Americans with modest annual incomes ($50 – $100,000) is much less than its impact on the working poor in all cultures. The consequences of rising prices are of most significance in Third World communities, where social circumstances make it difficult to maintain self-sufficient economies that are not dependent on global trading networks. In the early stages of biocatastrophe, massive food aid provided by well-to-do nations and non-governmental organizations (NGOs) such as the UN World Health Organization (WHO), Feed the Children, and Oxfam, may be sufficient to feed a small percentage of the rapidly growing population suffering from food insecurity (and lack of potable water, health care, and employment opportunities) for a period of a few years or a decade. As the world food crisis becomes more severe, technological solutions to increasing Third World food productivity will become increasingly compromised by the rapid decline in the world's oceanic fisheries and agricultural productivity. As food stress increases, the declining resources of the world's NGOs will become inadequate to meet the growing needs of a world food crisis.

The Limiting Factor of Accelerating Commodities Prices

A recent focus of American and European mass media has been the rise in commodity prices, especially gasoline and food supplies and the impact of these increases on working and middle class families. Mass media, especially in the US, has yet to acknowledge that these commodity price increases are just the beginning of worldwide shortages in fuel and food supplies. Such shortages will quickly limit the lifestyle options and reduce the prosperity and relative annual income of middle class participants in the global consumer culture, many of whom have already not only spent all of their own assets, but have accumulated significant indebtedness under the guise of flag-waving, patriotic, free enterprise advocates. As a result, American consumer cultists, enthusiasts, and recidivists (read: you, I, and the middleclass) have made a significant contribution to the current world debt of 600 trillion dollars. The early stages of worldwide biocatastrophe will be characterized by the dramatic impact of commodity price increases and the increasing unavailability of these commodities to the growing populations of the un- and under-employed as well as to large populations of working and retired people, many of whom cannot, or will soon not be able, to afford their costs. The perpetual indebtedness of a large portion of the middle class citizens of global consumer society will be a major component of the evolution of future world financial crisis. The growing inability of global consumer society to continue to produce low cost food and fuel will be the primary limiting factor for future population growth and increases in gross domestic product (GDP) and will be one of several phenomena that will transform much of America's and the world's middle class into a stress-ridden underclass. These limiting factors will have a far greater social and economic impact than cataclysmic climate change, which will be a secondary, delayed cause of economic collapse, social disintegration, and psychological distress.

Invisible Economic Impact: Energy Resources

The impact of rising commodity prices is closely associated with the economic status of affected communities. In the early stages of biocatastrophe, the impact of the increasing cost of basic commodities, such as gasoline and fuel oil, for lower income individuals is nearly invisible to those in the social milieu of upper income level communities. The higher the price of energy, the smaller the population that can afford to buy it. Increasing worldwide demand for energy from all sources is now rapidly depleting stocks of easily retrievable non-renewable fossil fuels. Other non-renewable fossil fuel stocks are plentiful, but difficult and expensive to extract. In the case of hydrofracking to recover natural gas in shale rock formations, the environmental costs of contaminated aquifers and surface water sources that result from this process are enormous. The use of hydrofracking to recover natural gas from shale rock formations is one example of the high external costs – in this case, contaminated nonrenewable aquifers – of our dependence on fossil fuels. Even more tragic are the deleterious economic consequences of the ongoing Gulf oil spill disaster. As biocatastrophe impacts a larger percentage of the world's population, only an increasingly small percentage of well-to-do consumers will be able to afford sophisticated, alternative, sustainable energy systems, healthy food, and potable water supplies, at least in the early stages of biocatastrophe. The speculative component of energy supplies as a commodity is graphically illustrated by the erratic rise and fall of gasoline prices before and during the global financial crisis of 2008-?. Erratic increases and declines in the price of oil due to the activities of predatory commodities speculators, including hedge funds and commercial banks, such as Goldman Sachs and Morgan Stanley, further undermine fuel price stability. Worsening economic conditions will exacerbate the relative cost of fuel and other basic commodities for a large percentage of the world's population already experiencing food and fresh water insecurity, underemployment or unemployment, falling incomes, and increasing anxiety about future economic opportunities for their children.

The Problem of Equitable Food Distribution

A well known challenge of our now globalized "free enterprise" market economy is the equitable distribution of basic food commodities to all world citizens. Many of the limiting factors of food availability involve future restrictions and dislocations due to climate change and declining ecosystems productivity and will occur over a period of time – possibly decades. Many limiting factors will result from the impact of climate change feedback mechanisms; others from ecosystem destruction and contamination. Presently, (2010) industrial agricultural systems are very efficient at mass producing huge quantities of food. Leaving aside the many quality and external cost issues, current food production is more than sufficient to feed the world's population now and possibly for the next decade. Unfortunately, as the earthquake in Haiti dramatically illustrated, food stress traditionally associated with third world populations since the first intrusions of predatory colonial western market economies beginning in the 16th century is now rapidly accelerating due to a combination of widening world income disparities, rapidly increasing transportation costs, and the logistical challenges of getting food stocks to those in need. Media coverage focus on the politics of Wall Street and the post TARP recovery of the stock market and the huge bonuses of early 2010 obscure the continuing saga of the continuing unavailability of food for almost one half of the world's population – especially those two billion people who lived on two dollars a day and often much less. They cannot afford the costs of the fossil fuels, especially oil, needed to transport industrial agricultural produce to their communities. Presently, first world governments and NGO assistance helps mitigate a small component of the world food crisis. In the future, the food availability disparities between the beneficiaries of our "flat-world" digital economy and the many citizens who live on its periphery will only increase.

Global Market Economy

Biocatastrophe occurs in the context of the ongoing collapse of a global economy that redistributes wealth to a few while constricting economic opportunities for most middle class world citizens dependent on the financial and commercial supply networks of its infrastructure. Population groups best able to survive the economic and social dislocations of a collapsing global economy will be rural communities with sustainable economies and a minimal dependence on fossil fuels and the complex trading networks of a global economy. Pockets of innovated techno-elite entrepreneurs, essential health care providers, and omnipresent state security apparatus and other governmental workers may be the last to be impacted by a collapsing global economy. The most optimistic take on future economic developments is that the world financial crisis of 2008-? may be a prelude to the reemergence of a more sustainable global economy in the future. The length of any such recovery is uncertain and unknowable; the current world financial crisis may also be an unsettling prelude to what will be a vastly larger collapse of the global market economy of the future. Such an event will be characterized by the declining availability of the financial resources needed to mitigate the social and economic consequences of human ecosystems collapse. Families, communities, corporations, and governments facing the challenges of the future will already be stressed by decades of debt accumulation and budget deficits. The synergistic interaction of the social, economic, and environmental factors that underlie the onset of the age of biocatastrophe will be characterized by the continued loss of both ecosystem biodiversity and productivity, and the collapse of non-sustainable, nonessential human ecosystems. Will the quiet tree-lined streets of Wellesley Hills meet the same fate as the vast expanses of America's toxic suburban landscapes if the global financial crisis of 2008-? becomes the global financial catastrophe of 2045? Or 2063?

The Social Dynamics of Increasing Unemployment

Disruptions in the global market economy have a potential to create self-sustaining increases in unemployment and underemployment. The first victims of anomalies in world market economies are the populations of Third World, the uneducated, urban and

80

rural poor who are already experiencing the impact of declining employment opportunities. Disruptions in local and regional agricultural ecosystems by industrial agriculture force large numbers of people into urban areas, where they become dependent on a highly technometabolic but erratic global economy. As increasing debt and financial uncertainty undermine the productivity of the global market economy, unemployment and underemployment spread to western economies and gradually extend to middle and upper income individuals. Increasing unemployment and decreasing productivity and incomes exacerbate the potential for global financial market collapse. Unemployment rates of 5-10%, which characterize the early years of an expanding global market economy, have the potential to become endemic, growing to more than a 50% unemployment or underemployment rate, increasingly impacting the well educated, formerly middle class communities that make the global market economy run on time.

Public Resources

Biocatastrophe is characterized by an increasing scarcity of local, regional, federal, and international government revenues i.e. accelerating needs and declining resources. As a growing global market economy begins its decline, fewer workers will be earning ever diminishing incomes and pay less taxes to ever more challenged governments. NGOs may temporarily play an increasingly larger role in mitigating the impact of biocatastrophe on some communities. In most western economies, the efficiency of a productive workforce essential for the operation of hospitals, transportation systems, communications, public safety, and food and water supply systems will be compromised by economic collapse. At some future point in time, governmental entities may no longer be able to supply the food and water needed by the victims of increasingly intense natural disasters in a world of an unreliable global market economy subject to sudden recessions, depressions, and collapse. Fewer public resources, including elusive tax dollars, will be available to meet the increasing needs of the growing populations of individuals and communities impacted by biocatastrophe.

Ecotoxins and Debt as Assets

The world financial crisis that began in September 2008 served the unique function of disclosing how much the prosperity of the era of Reaganomics depended on the accumulation of debt and the lucrative payment of interest and fees on financial instruments based on the sale of debt as an asset. Less obvious, but lurking in the shadows of the global financial crisis, is the long established propensity of western market economies, especially the American free enterprise system, to ignore the existence of ecotoxins as a form of entropy and the legacy of the technometabolism of a global consumer society. The short term benefits of the agricultural use of pesticides containing persistent organic pollutants (POPs) are only one example of the anthropological tendency to manipulate, exploit, contaminate, or consume natural ecosystems for the purpose of financial gain. The eventual costs of the consequences of the use of pesticides, herbicides, and genetically modified organisms to facilitate expanded world food production will become an unfunded legacy cost for future generations. The rapid expansion in world debt since the beginning of the Reaganomics era (1981) by western market economies and an ever growing global consumer society replicate the historical quest for natural resources (precious metals, fish, forest products, etc.) as the assets of public and/or private for-profit entities. The illusion that the profitability and productivity of intensely technometabolic economies (ever increasing growth based on the ever increasing expenditure of energy) does not include entropy (pollution, indebtedness, disorder, energy expended and thus not available) as a natural and unavoidable component of ecology of money ecosystems is a defining characteristic of our denial of the phenomenon of biocatastrophe.

Wall St. Ecotoxins

Massive petrochemical production of potent POPs, EDCs, and other ecotoxins in the form of pesticides, herbicides, pharmaceuticals, the endless variety of consumer product-derived ecotoxins, and potent global warfare-related toxic chemicals have resulted in decades of huge profits for the companies, stockholders, bankers, and Wall St. traders who facilitated their production and sale. The profit derived from the corporate manufacture, sale, and distribution of Wall St. ecotoxins, often as a component of the waging of global warfare, represents the energy derived from these products. The pollution, adverse human health impact, and ecosystems destruction constitute the entropy that is the consequence of these for-profit activities. The worldwide manufacture of these mostly synthetic chemicals, almost all of which originate from fossil fuel-derived petrochemicals, has resulted in the proliferation of a multiplicity of global ecotoxin contaminant pulses, which can now be measured in all biotic and abiotic media. In a chemosphere with limited natural resources, the profits to be made from the ecology of money are as finite as its entropic consequences are infinite. The ongoing annihilation of many species and ecosystems is due in part to the for-profit production of these socially useful poisons, i.e. the legacy of Wall Street ecotoxins.

Technometabolism

As the phenomenon of biocatastrophe spreads through urban, suburban, and rural communities of all income levels in the coming decades, the energetic technometabolism characterizing what is now a growing population of information technology beneficiaries will be increasingly restricted to populations in upper income brackets and increasingly isolated pockets of information technology entrepreneurs who are able to maintain access to whole foods and renewable energy resources. Peak levels of consumer culture consumption may already have been reached in the developed countries of the west, including the U.S. Future declines in conspicuous consumption may be paralleled by a florescence of electronic and genomic technologies as tuned-in flat-world information age participants fight for survival. If highly technometabolic, sustainable economies evolve as a footnote to the collapse of global military/industrial/consumer society, their prime mover will be electronic equipment and the florescence of information technology that is already the prime mover of society in the early years of the age of biocatastrophe. Their success will depend in part on meeting the challenges of the ecotoxic legacy of silicon chip production, electronic equipment and consumer product manufacturing, and the emerging nanotechnologies of post-industrial society. An unresolved question is the extent to which future income disparities will be reduced between sustainable techno-elite communities and the vast majority of the working and rural poor who will be the essential facilitators of their successful functioning.

The Downside of the World is Flat

When Thomas L. Friedman wrote *The World is Flat 3.0* (2006), at the peak of the post-911 real estate and stock market bubble, he was expressing an optimism about globalization and the proliferation of opportunities for most or all world citizens to share in the benefits of a western market economy fueled by the efficiency and ease of communication facilitated by information technologies. Citizens in any nation, and especially those in developing countries, upon achieving a modest level of education and finesse at operating electronic equipment, could participate in what, by the early 21st century, was a rapidly expanding global consumer culture. What Friedman failed to acknowledge or discuss was the obvious downside of flat-world electronic technology: the use of electronic technology by an ever increasing number of rapidly growing multi-national corporations to gain control of and manipulate commodities and consumer product markets throughout the world – in developed, developing, and third world countries. This spreading flat-world control began in the 1980s and coincided with the introduction of digital technology and the continuing success of industrial agriculture. The downside of this success was its predatory use of potent pesticides, fossil fuel-devouring fertilizers, and the destruction of local and regional sustainable agricultural communities. By the time Friedman completed his book, multi-national corporate control of flat-world information technology-based industries had spread to every component of global consumer society, ranging from McDonalds in China to farm raised swordfish from Indonesia. Particularly notable was the manufacturing of the electronic equipment invented and designed by America's techno-elite in the growing economies of developing nations, and their implementation via outsourcing by the millions of residents of these countries who replaced the now obsolete manufacturing jobs in America. By 2006, the benefits of flat-world technology now included the growth hormones and genetically modified seed varieties that extended the temporary hegemony of industrial agriculture to all corners of the world; examples of the proliferation of new technologies with ever-growing unanticipated downsides.

Economics 103

In Europe, America, and the growing cities of Asia, a vast middle class now has incomes of $35 to $100 a day. These are the workers in the sprawling infrastructure of global consumer society. Teachers in Egypt, lorry drivers in Scotland, sales personnel in retail stores in many nations, accountants in Singapore, clerks in the vast bureaucracies of governments and corporations, the anonymous workers who install dry wall in Paris or water pipes in London, luggage handlers at airports, the ubiquitous security guard; all are the essential workers that keep the infrastructure of global consumer society functioning. If they cannot afford rent, gas, groceries, if they are dislocated because of regional sectarian warfare, or if they are victims of ABRBs, or other newly emerging or reemerging pathogens, how can they function as workers in the global economy? How secure are their jobs in the context of financial depression and unemployment? What is the hidden impact of the biomagnification of ecotoxins on their health, that of their children, and their ability to function in a world of increasingly stressed economic and social forces? As the economic, social, and health physics impact of biocatastrophe increases over a period of decades, anthropogenic ecosystem collapse will victimize the workers of the global economy, many of whom are already stressed by the ongoing global financial crises that began in 2008.

U.S. Real Estate Mortgage Crisis

The proliferation of subprime mortgages and their repackaging and sale by predatory American financial engineers and institutions was the catalyst for an American banking crisis, which began in late 2007. Essentially a painful, prolonged epiphany as well as a prelude to a more extensive world financial crisis, the subprime mortgage crisis revealed that overvalued real estate was owned and often refinanced by individuals without an enough assets to secure the loans. The huge unsecured real estate debts were then repackaged and resold as highly leveraged securities called collateralized debt obligations (CDOs), leading to the development of a highly leveraged global economy where the total debt of financial institutions (corporate, real estate, stock and bonds, consumer debt) and the legacy debts of American and other governments are now at least twice as large as total world assets. Worldwide global market economy indebtedness and its vulnerability to sudden collapse illustrate the limiting factor of the ecology of money. All social classes experience the economic effects of biocatastrophe, even in its early stages.

The Ponzi Schemes of the Shadow Banking System

The multiple scams underlying the global financial crisis of 2008-? were engendered by several decades of speculative real estate investments, ballooning corporate debt, and the excessive accumulation of consumer product, automobile, credit card, and college education debt. The core of the current global financial crisis derives not only from the American subprime real estate disaster (an excuse and a diversion as well as a trigger), but also from overvalued corporate stocks, bonds, and other assets, as illustrated by the Enron debacle. Out of 140 trillion dollars in current world assets (business property, stocks, bonds, and other paper conveyances) (Farrell 2005), a reasonable estimate is that at least 25% of these assets are secured by unleveraged debt, often insured by the credit default swaps of our shadow banking system. In the Reaganomics era of multiple shadow banking Ponzi schemes, these highly leveraged debts were packaged and resold as interest paying "assets". The potential loss of these assets was insured in the form of credit default swaps (CDSs) paying significant interest rates to the asset holder until such time as the debt was paid by the borrower. This ingenious Ponzi scheme, concocted by the enthusiastic financial engineers of the American Environmental Fascist Free Enterprise System (AEFFES), was highly remunerative until such time as leveraged debt became so great the entire system collapsed. The ultimate victims, often foreign banks, governments, and wealthy investors, were the unwitting buyers of these highly rated securities, which are actually repackaged mortgages and corporate debt. A large proportion of these allegedly four star quality investments are or will soon be worthless debt. In the world financial crisis of 2008-?, important commercial borrowers that could not afford to pay these debts were subsidized by governmental bailouts. Small investors, especially those with 401(k) plans who invested in stocks whose values plunged during the financial crisis, and many of the original mortgage holders, have been unable to secure bailout funding and have been victims of the snowballing financial collapse. The cumulative effect of falling real estate values and rising unemployment has further undermined the value of real estate held by middle class families that was not initially in the subprime category. As the global financial crisis of 2008-? unfolds, private investors and financial institutions are increasing unable to pay the ±24 trillion dollars of interest due on world debt, let alone the debt itself. In both the US and Europe, increasing socialization (nationalization) of now bankrupt financial institutions is the only alternative to the total, if premature, collapse of global financial markets. In the future, government entities will not be able to print enough money to pay accumulated free market debts, and global and regional financial ecosystems founded on the philosophy of economic growth, the extension of unsecured credit, and tax cuts for the palisaded elite, will collapse.

The Global Banking Crisis of 2008-?

The unexpected collapse of first the American, and then, the international network of investment and commercial banking institutions, is an omen of much more serious potential infrastructure collapse that will likely accompany full-fledged biocatastrophe. The surprise element is this crisis was the domino effect of overdue US residential real estate mortgage-based debt on a commercial banking network burdened with corporate debt, commercial real estate debt, stock, bond, and derivatives debt, and credit card, auto loan, consumer product, and student loan debt. The Medicaid, Medicare, pension funding, TARP bailout, and Freddie-Mac and Fannie-Mae legacy commitments are the storm clouds on the horizon line of private and corporate debt. Financial systems collapse may precede expected pandemics, commodities shortages, and the worldwide unavailability of fresh water and whole foods. The initial socialization of a small component of accumulated global banking debt following the October 2008 crisis by the United States and European governments may only be a stopgap measure postponing eventual global financial collapse. The inevitable collapse of growth-based western consumer society will be an integral component of the phenomenon of biocatastrophe. Given the time frame of the evolution of biocatastrophe, it is reasonable to project this collapse as occurring between 2025 and 2075. The surprising elements in the unexpected financial crisis of 2008-? are its severity and the possibility that it represents the early stage of the economic collapse of an unsustainable world economy. The hopeful element in the current global financial crisis is that a return to a sustainable global economy, one not dependent on the accelerating accumulation of debt, can be achieved in painful stages in the next few decades as an alternative to the specter of global financial collapse.

Changing Landscapes

In London at 5:00 PM, vast crowds of neck-tied bankers congregate at pubs, though perhaps in lesser numbers after the collapse of Lehman Brothers in September 2008. In America, large numbers of commuters endure the torture of massive traffic jams from the quiet back roads of New England suburbia to the freeways of Los Angeles. The invasive sprawl of suburbia spreads across the desert landscape of the American southwest and the denuded landscapes of overpopulated Florida like a slime mold. China, India, and Indonesia have their growing hordes of upwardly mobile commuters, telemarketers, and world traders, whose incomes are proportionally higher than the remuneration of the workers who make the success of the global industrial-consumer society possible. How do the relatively urban, electronically sophisticated, educated

86

elite respond to shortages of basic commodities and fresh water, growing worldwide malnutrition, and ecotoxin contaminant signals in the foods grown or bred in all areas of the world? How many of their children will suffer from autism spectrum disorders, obesity, diabetes, ADHD, and the spread and magnification of neurotoxins in their daily diet? How will the stressed lifestyles of the working class so essential to the functioning of global consumer society impact the necessary efficiency of its vulnerable interconnected networks of technometabolic ecosystems (communication, transportation, health care, public safety, etc.) as biocatastrophe becomes an unnerving reality instead of a theoretical possibility?

The History of Asset Growth

When that robust maritime economy of colonial New England and the early Republic (1640 – 1860) traded codfish, clapboards, house frames, the products of the cooper, and rum from now forgotten fishing ports and inland mill towns as well as the ports of Boston and Salem, for Swedish iron, molasses, cotton, and the ceramics of the East India trade, their economy had a favorable balance of payments. Biosphere was bank account. With the help of the neutral trade (1796-1807), the slaves of the Caribbean who harvested the sugar cane used to make rum, and the wealth derived from the tropical hardwoods of the Bay of Campeche, New Englanders and others built the wooden ships that were the basis for economic growth, as well as those stately federal houses we still see in Newport, Boston, Salem, Newburyport, MA, Portsmouth, NH, and Wiscasset, ME. The robust period of America's industrial growth that followed was fueled by the unlimited horizons of a westward expanding economy, at least until the advent of global warfare and depression changed the world we live in during the 20th century. For two centuries, our collective assets and our balance of trade had been far in excess of our meager debts. At some point in the 20th century, in a world of competition for the diminishing assets of a biosphere of finite resources, national (and world) debts began accumulating. Sometime between the death of John Kennedy (1963) and the beginning of the era of Reaganomics the growth in our collective indebtedness began catching up with and then exceeding the growth in our assets. Non-sustainable global military/industrial/consumer society (GMICS) was born. Few writers or commentators acknowledged the increasingly obvious non-sustainable elements within our wonderful consumer culture that changed the orange groves of Largo, FL, into one huge parking lot, or covered large portions of the desert southwest with a growing masonry slime mold. Now many of these same areas are suffering the indignities of the subprime mortgage crisis, rapidly increasing unemployment, and the growing water stress that will soon be a major limiting factor for the growth of suburbia in many states. The rapid growth of indebtedness in the age of chemical fallout to fund the illusion of a vast increase in assets laid the groundwork for the birth of biocatastrophe. The global financial crisis of 2008-? represents the infancy of the age of biocatastrophe, an age in which any future economic growth is accompanied by an equal, if not greater, expansion of world indebtedness. At this late stage in the epic of human civilization, economic growth is dependent on the growth of world debt. Nobody, not even the new Obama administration, nor important media that keep us informed on such issues, is yet ready to break the bad news about our current predicament, i.e. any future economic growth can only be accompanied by the incursion of further indebtedness. The western model of an ever growing market economy has long passed the inexorable boundary of sustainability.

Debt versus Assets

One of the major problems bedeviling modern western industrial consumer society is the relationship between world debt and world assets. Total world debt is now approaching at least 600 trillion dollars and growing rapidly (Ferguson 2008). This figure includes the obligations of US and other international governments to fund the legacy costs of pension funds, Social Security, Medicare, Medicaid, and other social contracts, and the totality of real estate, banking, and institutional investment debts, stock and bond trading debts, and corporate and personal debt held by all the participants in global consumer society. Total world indebtedness may be compared to a rough sketch of world assets using the slightly exaggerated figure of 7 billion people as the world's current population. The middle column of the following table is the median of assets held by each population category. Far more than a half a billion people have zero assets; many middle class participants in the global consumer society have a total indebtedness (real estate, credit cards, automobile, and college loans), which now far exceeds their assets. But this is a sketch, not a world census bureau report.

Assets sketch:

0.5 billion people	$0 assets	
3.5 billion people	$1,000 each	3.5 trillion ($3.5 \ 10^9 \times 10^3$)
3 billion people The working middle class	$10,000 each	30 trillion ($3.0^9 \times 10^4 = 30^{12}$)
50 million people The palisaded elite	$1,000,000 each	50 trillion ($50^6 \times 10^6 = 50^{12}$)

Of those 50 million people, some have additional assets:

1 million people (the rich and the beautiful)	$16.5 million	16.5 trillion ($1^6 \times 16.5^6 = 16.5^{12}$)
100,000 people (super-rich industrialists etc.)	$100 million	10 trillion ($10^5 \times 10^8 = 10^{13}$)
Misc. world assets	----	90 trillion
Total world assets		200 trillion

While the total amounts of world debt and world assets are certainly debatable, our sketch indicates that world debts are now three times world assets; this, in fact, may be relatively optimistic interpretation of the current situation. This raises the annoying question of who is going to pay these debts. Even more perplexing is the challenge faced by the growing millions of middle class participants in the global consumer culture whose assets are suddenly worth far less than their indebtedness, and whose monthly living expenses are far more than their income. What now?

Debt Payment?

The current world financial crisis has been exacerbated by a combination of the growing inability of mortgage holders, investors, and the shadow banking network to pay the annual interest due on world debt and the inscrutable paperwork labyrinth invented by the financial engineers who created the collateralized debt obligations (CDOs) that evolved from the repackaging and resale of subprime real estate mortgages and other debts as assets. The credit default swaps (CDSs) created to insure these phony assets result in highly profitable interest payments and commissions, which must be paid on these overvalued assets. Once an "insured" asset declines in value, insurance payments become due, triggering the cascade of additional asset value declines now the core of the global financial crisis. A rough sketch of the amount of interest to be collected on a yearly basis by commercial banks, investors, bond holders, and "credit default swap" enthusiasts, i.e. shadow bankers, on world debt is ±24 trillion dollars per year. Almost all of the current world debt has been incurred since the beginning of the Age of Reaganomics (1981). Until the advent of the global financial crisis, world banks, including the recently evolved shadow banking network, have reaped the benefits of collecting the interest due on growing world indebtedness for 2½ decades. This profitable enterprise has resulted in the construction of thousands of shiny skyscrapers symbolic of the illusion of growing world wealth. Recent bonuses paid to New York bankers as commissions on the manipulation of these debts via stock trading, mortgages, bonds, derivatives, and the now infamous credit default swaps have been in the range of hundreds of billions of dollars per year not including stock options, payroll, or trading company profits. Merrill Lynch has received wide publicity for paying 15 billion dollars in bonuses just prior to its government sponsored takeover by bailed out Bank of America, an amount significantly greater than America's payment of food aid to starving communities in Africa. Financing the world's debt has been an extremely lucrative profession, not only on Wall Street, but in England, Scotland, Germany, France, and Switzerland and other centers of world banking consortiums, at least until the recent collapse of this Ponzi scheme. In this context, if the federal government socializes part of American banking losses due to their investment in world debt (let's just estimate that American banks hold 100 trillion of the 600 trillion of world debt, a very conservative estimate,) with TARP-type government bailouts and guaranteed loans of, say, 5 trillion dollars per year, will this really solve the problem faced by our insolvent banking system?

George Soros on the Financial Panic of 2008/09

There is the possibility that the ongoing collapse of the world financial systems, rather than being a temporary, severe recession or a depression, may be the first stage in the collapse of a non-sustainable, highly indebted, global market economy. Financial system recovery to a more sustainable equilibrium may occur in a year or two, or possibly not at all. George Soros, the infamous hedge fund king, provides an interesting perspective on this crisis but makes no connection between the financial crisis and the growing world environmental crisis. "Financial engineering reached such heights of complexity that the regulators could no longer calculate the risks… Financial capital started fleeing from the periphery to the center… Commodity prices dropped like a stone and interest rates in emerging markets soared. So did premiums on insurance against credit default. Hedge funds and other leveraged investors suffered enormous losses precipitating margin calls and forced selling that has also spread to markets at the center…The issuers of commercial paper were forced to draw down their credit lines, bringing interbank lending to a standstill…The race to save the international financial system is still ongoing." (Soros, 2008b).

George Soros Recapitulated

George Soros, along with other important commentators like Paul Krugman, Nial Ferguson, and Freeman Dyson provide insightful commentary on economic and environmental issues in many forums. Soros, recently called the hedge fund king (NYRB – Dec 2008) is an extremely successful investor. Let's speculate that George's net worth is 3 billion dollars, placing him among the top 1,000 of the world's most wealthy individuals. Of most interest is that while Soros holds the 3 billion in assets, (admittedly paper assets, that most ephemeral form of wealth,) some other group of institutions or individuals now hold that 3 billion dollars as debt, given the growing imbalance between rising world debt and falling world asset values. In Soros' case, at least a billion dollars was derived from insightful hedge fund betting with respect to the devaluation of the British pound in the 1990s. If investors somewhere are collecting say 4% interest on that billion dollar debt, 40 million dollars per year is being added to world indebtedness by the investors in the world shadow banking network who profit from the interest payments on that debt. In terms of world indebtedness, the very nature of the success of the predatory shadow banking system that grew out of the deregulation of the narcissistic do-as-you-please era of Reaganomics was that, before the collapse of this unsustainable financial Ponzi scheme, a world debt of 600 trillion dollars at 4% interest would yield 24 trillion dollars a year in interest payments. This helps explain why financial institutions in all western countries, led by America's predatory banking and investment system (the veritable stupids,) were so anxious to create the collateralized debt obligations and other phony derivatives and the credit default swaps that were becoming so profitable to formulate and market. Soros and others used the term "financial engineering" as a description of this, until recently, very lucrative profession. Unfortunately, we now have to borrow the 24 trillion dollars needed per year to pay the interest on our collective debt, which has evolved from a quarter century of unregulated financial engineering. This represents an annual payment of $3,400 per person per year, utilizing our hang loose world population estimate of 7 billion people. Who was the (nit) wit who suggested we, as a collective global consumer society, were going to grow our way out of debt?

A Silver Lining to the Financial Collapse?

The worldwide collapse of a global financial system based on the illusions of the model of an American free enterprise market economy (read: AEFFES) may have the hidden benefit of slowing the accelerating discharge of greenhouse gases and ecotoxins into the environment. A decline or even a slowdown in world gross economic activity ultimately entails a decline in the production of both CO_2 and a wide variety of ecotoxins and a slower increase in the accelerating rate of the continuing degradation of the biosphere. Slower growth in China in particular means fewer coal-fired plants and thus lower levels mercuric sulfide emissions. A slowdown in consumer product production, especially of electronic equipment and its plethora of ecotoxic wastes, means less PBDEs, PFCs, and heavy metals will enter the environment. Unfortunately, even if all industrial/consumer products and transportation-related ecotoxic emissions are completely halted, given the non-retrievable quantities of greenhouse gases and anthropogenic ecotoxins emitted since 1940, the fundamental biodynamics of biocatastrophe are irreversible. The unresolved question is its future timeline and the sequences of events that will accompany its maturation.

Afterthought

The only good news is that an early start to the global financial crisis implicit in the dynamics of biocatastrophe nurtures the hope that a painful return to sustainable economies may yet be implemented as an alternative to global social and economic collapse. Given the realities of cataclysmic climate change, ecotoxins in pathways to human consumption, and ecocatastrophe on a global scale, any alternative to total global financial collapse may be wishful thinking. But humankind is both innovative and optimistic. Maybe in 2200, Paris will still be Paris but New York City? With respect to the upcoming Netflix series "The Nightmares of Wall Street" the thought arises that catastrophic sea level rise may have its benefits. Unfortunately, when the New York subway system is destroyed by a category 5 hurricane (or even a category 3 storm?) the predatory shadow bankers of Wall Street can flee to St. Moritz, various Caribbean islands, or a penthouse in Dubai. The working citizens of New York City, many of whom have already had many of their assets "skimmed" by shadow banking predators, will be the victims of this likely "unexpected" washover event.

XIII. Biocatastrophe as the Implosion of Human Ecosystems

Denial of Ecological Disaster

By 2010, a growing community of American citizens will join the large minority of European citizens in an attempt to limit some of their CO_2 emissions by adopting renewable energy sources or otherwise reducing their carbon footprint. What had been the concern of a small minority of environmentalists is becoming a popular movement among a much larger and much more diverse community of individuals seeking sane sustainable lifestyles, a movement given a huge boost by the newly elected Obama administration. The highly visible phenomenon of melting glaciers and polar ice is indicative of the long term threats of rising sea levels and cataclysmic climate change now the focus of worldwide concern. Implicit in cataclysmic climate change is the certainty of increasingly intense droughts, hurricanes, and declining food supplies for populations in all areas of the globe. The post election acknowledgement of the problems posed by greenhouse gas emissions and global warning by the newly elected Obama administration is a welcome sign of changing attitudes, but neither political leaders nor the mass media have yet addressed the problem of the synergistic impact of a complicated series of environmental, financial, social, and political crises. The probability of what will be an increasingly rapid, nearly uncontrollable collapse of the non-sustainable components of global military/industrial/consumer society (GMICS) continues to be the subject of an ongoing and intense denial by America's mass media and the vast majority of its population. The incipient stage of the collapse of global consumer society may already be at hand. Unfortunately our growing financial indebtedness diverts attention and resources from the larger human predicament of how increasingly complex and technometabolic human ecosystems can survive in a biosphere of rapidly dwindling resources. The greatest challenge of this conundrum is how to mitigate growing income, health care, employment opportunity, and other disparities between the flourishing elite 5% of the world's population and the vast majority of the world's citizens who are rapidly falling behind in the context of a non-sustainable global consumer culture.

The Limiting Factor of Technology

The 20th century devotion to technological solutions to facilitate the urgent need to promote the "growth" and thus the expansion of market economies, including the hang-loose American free enterprise system or the more tightly controlled socialized European and Asian market economies, is expressed as the increasingly intense technometabolism of the prime movers of western global military/industrial/consumer society. The ease and convenience of steam engines (read nuclear power plant turbines, etc.,) electric power grids, internal combustion engines, petrochemical plants, and weapons production has resulted in the evolution of a complex global industrial consumer culture characterized by extraordinary advances in electronic communications systems, including computers, the internet, and personal electronic devices. These advances obscure the fact that implicit in the ecology of technology is that the benefits of the expenditure of energy precede the later appearance of entropy (energy spent and not again available). All technologies have their hidden costs, as in the greenhouse gas emissions of internal combustion engines, or the legacy of polybrominated diphenyl ethers (PBDEs) and other ecotoxins created during the production of modern electronic equipment. The limiting factor of entropy is the legacy of technometabolism and its depletion and destruction of ecosystem viability, biodiversity, and productivity.

To Green or Not to Green

When Thomas L. Friedman published his *Hot, Flat, and Crowded* in 2008, written before the collapse of Lehman Brothers, a tone of urgency dominated his delineation of the necessity for a new paradigm for global consumer culture in the future, which he expressed in his subtitle "Why we need a green revolution – and how it can renew America." Implicit in this treatise is the acknowledgement of the downside of flat-world information technology: its long term impact on facilitating greenhouse gas emissions, global warming, and cataclysmic climate change. The intellectual, political, and economic dynamics of meeting this challenge combine the efficiency and efficacy of electronic devices with the ability of flat-world technology beneficiaries to facilitate this green revolution. In the context of a globalized world of ever more powerful multinational corporations, including a long entrenched global oil industry, growing and highly sectarian fundamentalist religious communities, and an increasingly marginalized academic elite (including greens), Friedman does not make a compelling case that there is a sufficiently powerful and large enough pool of concerned citizens to facilitate a green revolution given the ongoing browning and increasing domestic unrest of "post-American" society (Zachariah 2008).

Visibility vs. Invisibility

One may watch the onset of biocatastrophe on TV: the global financial crisis of 2008-?, the world food crisis, increasing commodities prices, declining fresh water supplies, global warming, increasing hurricane intensity, melting glacier and polar ice, devastating brush fires in Australia, declining bee, bat, and polar bear populations, and the spectacular Gulf oil spill disaster. Resource depletion is obvious, painfully manifested in higher gas and food prices, the obscene destruction of biodiversity in most ecosystems, the worldwide production of biofuels instead of food in gradually

94

shrinking agricultural regions, deforestation, and the loss of productivity in oceanic fisheries. In contrast, no specific ecotoxin, with the possible exception of methylmercury, is present in most biotic media in sufficient concentrations to cause observable health effects. The synergistic impact of thousands of ecotoxins, including hundreds of neurotoxins, initially present in the global atmospheric water cycle at very low concentration levels, causes highly visible health effects, including cancer, asthma, impotency, birth defects, immunological disorders, autism, and learning and behavioral disorders, when biomagnified at higher trophic levels or when ingested as ubiquitous endocrine disrupting chemicals (EDCs). None of these adverse health effects can be attributed to a specific ecotoxin. The link between multiple micro-doses of invisible ecotoxins and growing cancer rates is too subtle, too complicated, and, in fact, too controversial to be understood and accepted by the general public. The proliferation of multiple ecotoxins in pathways to human consumption impacts all social communities, but its role in the etiology of disease is much less visible and more easily rationalized than the consequences of cataclysmic climate change, resource depletion, the global financial crisis, or unemployment.

The Palisaded Elite

Accelerating global climate change, resource depletion, loss of biodiversity, the widening impact of chemical fallout ecotoxins, and the phenomena of antibiotic resistant and newly emerging/reemerging pathogens will be accompanied by the shrinkage of privileged communities with sufficient financial and material resources to avoid the social, health physics, and economic impacts and population declines felt by less privileged communities experiencing the consequences of non-sustainable agricultural, economic, and industrial activities. Palisaded estates, condominiums, or urban penthouses, even those inhabited by the innovative techno-elite or the fabulously rich and beautiful, may not provide protection from ecotoxins in food chains, pandemic events, social unrest, and/or disruptions in public utilities and transportation services during infrastructure collapse. A growing minority of these elite is and will be concerned about global climate change, global warfare, proliferation of ecotoxins, ecosystem collapse, and the impact of increasing commodity prices, food shortages, and the lack of fresh water on less fortunate communities. In America, the world leader in the evolution of narcissistic global consumer culture, its frenzied consumption of non-renewable resources, and the invisible worldwide spread of petrochemical-derived ecotoxins, a large portion of the palisaded elite will continue to be oblivious to the consequences of these developments. As the principal beneficiaries of the spread of global consumer culture, the need to maintain physical security in a world of social unrest and increasing economic uncertainties will dominate lifestyles and political attitudes even more than it does at present. In fortress America (and elsewhere), where the rich get richer and the middle class become the underclass, those powerful SUVs, with well armed security guards wearing sunglasses, are a harbinger of the early stages infrastructure collapse.

The Culture of Violence and the Power of Weapons

A key element in the evolution of the predatory narcissistic consumer culture denoted here with the polemical label American Environmental Fascist Free Enterprise System (AEFFES) is the culture of violence so often the unspoken context of the vitriolic commentary of American right wing media pundits. Perhaps more accurately described as a liberal cult of the gun (no regulation, do as you please, wave the flag and fire away), its mythology joins the inalienable right to bear arms in the US constitution and the ideology of the "chosen people" of evangelical Christianity with the self-indulgent free enterprise philosophy of conspicuous consumption and the perpetual accumulation of assets. The quest for wealth is closely associated with the cult of power and is often, but not always, associated with the veneration of firearms. "What comes around, goes around," is a commonplace truism with unfortunate consequences in a world of growing social unrest, jihad, spreading drug trafficking organizations (DTOs), ethnic civil wars, and sectarian violence. If American citizens have the inalienable right to bear arms of every caliber, so too do all world citizens, including thousands of teenage terrorists, many of whom share the hidden resentments and tendency towards violence of their American opponents. It will only take one or a few radicals or psychopaths with a surface to ground missile hidden in a camper to cause a LORCA by taking out a nuclear reactor water cooling system or destroying the internet with one well aimed missile directed at its vulnerable Silicon Valley center of operations. In the early years of the age of biocatastrophe, the venerated right to own rifles for hunting and small arms for self defense has morphed into the right to supply automatic weapons for drug trading organizations, to own powerful explosives, such as those used in IEDs (improvised explosive devices,) and culminates in the potential for the worldwide spread of sophisticated weapons, which will occur in proportion to the worldwide spread of economic collapse and social unrest.

Below the Limit of Perception and Detection

Two problems bedevil biological monitoring of ecotoxins in pathways to human consumption: one is political, the other scientific. One of the most significant scientific challenges with respect to chemical fallout ecotoxin contaminant pulses pertains to their measurement in the biosphere. Some ecotoxins, especially those associated with modern consumer products (BPA in plastics, PBDEs in electronic equipment) are directly ingested, inhaled as dust, or absorbed by humans in domestic and work environments. Most other ecotoxins have much more complex pathways to human consumption and are incorporated into the food chain by microorganisms such as bacteria and plankton, which ingest them as food and efficiently recycle them in the biogeochemical cycles that interconnect all food webs and ecosystems. They are then transported as contaminant signals in concentrations often ranging from ppt (parts per trillion) or less to ppb (parts per billion) in abiotic media (water) or biotic media on every trophic level. Once ingested by humans, the cross placental transfer of these ecotoxins, often resulting in genes with altered genetic codes due to exposure to EDCs, is the most important of all exposure pathways for the children who are the canaries in the coal mines of our great experiment with the effluents of global consumer culture. In most cases, these ecotoxins are initially present below the level of detection using current analytical technologies. Only after biomagnification in biotic media in higher trophic (feeding) levels, do most currently tracked ecotoxins become observable, usually at levels above 2 ng/g (nanograms per gram or ppb). Ecotoxins below this level of concentration seldom have observable adverse health effects. As multiple ecotoxins become biomagnified at higher trophic levels, accurate measurement with current technologies is only prevented by political considerations and a lack of governmental resources, which reflect the social values of a narcissistic consumer culture that is oblivious to genobiocide as an ongoing, inevitable, and accelerating tragedy of the World Commons. In this context, the dynamics of biocatastrophe are below the limits of social perception and, as yet, off-limits for political discussion. The massive biomonitoring reports of the Centers for Disease Control, some of which are cited in the *Bibliography* in *Volume 3*, as well as numerous academic and NGO reports, would seem to contradict this assertion. Unfortunately, given the ±100,000 chemicals now being produced and their proprietary status, the systematic biomonitoring of anthropogenic ecotoxins is just beginning.

An Urban Scenario

Infrastructure collapse will be manifested not only in financial markets but also in bathrooms: a shortage of toilet paper, soap, and service workers. Even a small outbreak of *C. difficile* among the urban working poor could quickly spread to subways, airports, and other public restrooms. A disruption in the supply of health care providers, sanitation workers, and sanitary supplies in an urban environment has the potential to result in the rapid spread of easily-transmittable ABRBs in rapid transit systems, in elevators and escalators, on stairway banisters, doorknobs, bathroom sinks and toilets, shower stalls, in school and workplace interiors, on cell phones, blackberries, computer keyboards, and into food service environments. The current health physics impact of the H1N1 virus, a relatively benign pandemic in contrast to what could occur in the future, is a predictor for the exposure pathways of future epidemics where hand to mouth contact is more important than the inhalation pathway. The worst case scenario in an urban environment in a *C. difficile* epidemic, for example, involves the unavailability of trained service workers to operate subways, airlines, and other essential services. Will there be Wall St. ecotoxin traders, or the rich and the beautiful, or just plain old mom, dad, and the kids, suddenly stuck in a tall apartment building in Manhattan? You say you have an old case of Lyme disease, your mother has MRSA, and you now have *C. difficile* and are too weak to walk down 16 flights of stairs? Furthermore, pizza delivery is not available tonight because the workers are ill with a new strain of Avian flu. What do you eat, where do you go, and who cleans your bathroom? And you say you are worried about global warming?

The Health Needs of Alpha Males

One of the ironies of the spread of ecotoxins through the food webs of the rich and the beautiful (read: respectable upper middle class families in Wellesley Hills, etc.) is that after pregnant mothers and their children, the next most vulnerable members of world society are educated, elite males in the higher income brackets (forget the invisible urban and rural poor of the world). As a rule, high energy alpha males feed at the highest trophic level of the food chain, and consume foods containing high levels of lipophilic (fat loving) ecotoxins, e.g. red meat and other fatty animal tissues or byproducts, blue fin tuna, swordfish, caviar, eggs, etc. Sophisticated pharmaceutical products (Viagra, Levitra, Cialis), designed to alleviate the consequences of the ingestion of these ecotoxins by alpha males (e.g. ED) are then excreted in waste treatment systems and migrate to drinking water supplies, and are subject to ingestion by humans or food chain uptake by microbial grazers. These pharmaceutical ecotoxins join a wide array of bovine growth hormones, antibiotics, and other pharmaceuticals already present in water cycles and food webs. Alpha males join children, mothers, and all world citizens as the potential victims of this chain of sequences. The onset of early puberty and other effects on teenage girls has already been demonstrated as a consequence of the inadvertent circulation and biomagnification of these pharmaceuticals and growth hormones. The sight of males walking on America's beaches with enlarged breasts is now as commonplace as Cape Cod horseshoe crabs are rare. The vast for-profit marketing of pharmaceuticals is a notable example of the careless distribution of consumer products that have a significant, if often invisible, delayed health impact on individuals who inadvertently ingest these ecotoxins as the accidental legacy of profit-driven corporate privateers. The ubiquitous presence and impact of pharmaceutical ecotoxins at lower levels of the food chain, symbolized by the feminization of male fish due to the intake of hormones such as steroid estrogens, is a forgotten footnote to the health needs of alpha males.

98

Mass Culture in the Age of Biocatastrophe

One of the unfortunate characteristics of global consumer society is the proliferation of processed foods with minimal nutrient value and maximum caloric, hydrogenated fat, food additive, and micro-ecotoxin content, in contrast to whole foods. The distribution of food and drinks, the main constituent of which is either processed flour or high fructose corn syrup, has been clearly linked to the increasing incidence of obesity and diabetes in western children. These processed foods, along with red meat and sea food, are important transport vectors for ecotoxins, including POPs, to both children and adults. In the coming age of commodity scarcity, a larger proportion of the world's population will be increasingly dependent on low cost, low quality, processed foods due to the high cost and/or lack of availability of healthy whole foods. Whole foods are also characterized by ecotoxin contaminant pulses as exemplified by methylmercury in seafood but remain nutritionally superior to processed foods. Implicit in the sociology of biocatastrophe is the fact that inhabitants of urban and lower income suburban and non food-producing rural communities will be the most vulnerable populations to the rising costs and unavailability of whole foods. For these populations, global warming and rising sea levels will be a distant and often irrelevant threat, at least until cataclysmic climate change triggers major food shortages or floods their local Wal-Mart super store.

Economics and Religious Dislocations and Disillusionment

A fundamental characteristic of American culture in the last decades of the 20th century, along with the appearance of the phenomenon of Reaganomics, was a decline in the traditional "main line" religions (Episcopalian, Methodist, Congregational, etc.) and their replacement by a surge of evangelical fundamentalism, cogently delineated by Phillips in *American Theocracy* (2006). A recent series of phenomena closely connected with the ongoing financial crisis, which manifested itself in the "near total meltdown" of September 2009, are the rapid increases in un- and under-employment in 2008 and 2009, the subprime mortgage crisis and subsequent flood of mortgage defaults, and the growing dissatisfaction with the ability of state and federal governments to ameliorate this economic downturn. While the manufacturing capacity of the United States has been in a long decline, the resulting loss of industrial economic productivity has been replaced by a financial services industry that now accounts for over 20% of gross national productivity. The rapidly increasing population of economically stressed, angry, lower and middle class workers join the evangelical fundamentalists in their dislike and suspicion of the Wall Street elite. All are aware of the rapidly increasing income disparities between the bankers who often earn more than a million dollars per year and the unemployed workers whose jobs have been made obsolete by efficient fiber optic information technologies. During the first year of the

99

Obama administration, characterized by a tortuous struggle to implement modest health care reforms, America's disenfranchised workers have joined evangelicals and other traditionally conservative elements of American society and coalesced into what is appropriately called Tea Party fascists. All have one thing in common, a hatred of the corporate elite, now symbolized by Wall Street shadow bankers, but also including the academic elite, the "left-leaning" writers of *The New York Times*, greens, such as Al Gore, and all members of Congress in the Democratic Party. This growing political schism ensures the least effective response to the accelerating implosion of the viability of American cultural and economic institutions.

An Uncertain Future

For those of us who live within the outer margins of an enclave of the palisaded elite, well away from the direct consequences of infrastructure collapse, urban epidemics, and the need for armed men (and women) driving Humvees, annoying questions keep intruding on uneventful daily routines - What will be the fate of our children's children and grandchildren? How vast will be the impact of the synergistic components of biocatastrophe? Will isolated communities be able to maintain sustainable lifestyles and creative economies? Will such communities survive the growing onslaught of ecotoxins in the food webs of the biosphere, or the threat of the proliferation of multiple ABRBs and GMOs? Who will tell our children and grandchildren we have a problem much greater than that posed by CO_2 emissions? How can we change our Environmental-fascist ethic of growth at any cost, the for-profit proliferation of Wall St. ecotoxins, the inevitable spread of ABRBs, and the possibility of pandemic diseases? What will be the world population in 2100? How many cultures and communities will weather the coming age of biocatastrophe and achieve equitable, sustainable economies based on renewable energy, sustainable local agriculture, and the use of renewable resources?

Sustainable Communities

The central issue in biocatastrophe is the extent to which population groups (nations: Iceland, Germany, New Zealand,) small communities with alternative lifestyles and creative economies (located within the USA or anywhere else,) or long-established rural communities with minimal dependence on global commodities markets can achieve sustainable economies. Will these nations or communities be able to maintain relative immunity from pandemics and water and food stress in a world of cataclysmic climate change, chemical fallout, ABRBs, GMOs, collapsing financial markets, spreading unemployment, declining asset values, commodity shortages, and social unrest? As we move from the dawn of the age of biocatastrophe to its full fledged presence (probably within 50 years), only a small percentage of the world's 7 to 10 billion people will remain unaffected by contaminated water, food insecurity, loss of biodiversity, the

100

spread of ecotoxins, ecosystem collapse, pandemics, the impact of GMOs, and the consequences of the collapse of a global economy based on the accumulation of ever-growing debt.

Biosphere as Bank Account

In the 16th century, incipient European market economies (Spain, The Netherlands, France, and England) began the systematic exploration and exploitation of the American continents. Biomass (codfish, beaver pelts, timber) and geomass (gold, silver, iron) became valuable components of the birth of a global market economy. Twentieth century ecosystems collapse is rooted in centuries of biosphere as bank account interface. The fundamental root of biocatastrophe is the imposition of anthropogenic ecosystems on natural ecosystems in the context of a predatory, sectarian free enterprise ethic. The propensity to cultivate or harvest natural resources evolved into the systematic exploitation of natural resources by profit-driven market economies. The survival of 8 billion people, including possibly as many as 3 billion future participants in the comfortable high-tech lifestyle of a global consumer culture, mandates the increasingly frenzied consumption of the finite resources of the biosphere. The systematic contamination of the global atmospheric water cycle by anthropogenic ecotoxins and the accumulation by the global economy of indebtedness equal to ± 10 times the gross world annual economic output are the natural outcomes of the quest of all races and creeds, not just white Christians, for the God given right to enjoy prosperity. A politically incorrect question to be asked at the dawn of the age of biocatastrophe is, what would God think of, as among one of the rights of man, his/her uncanny ability to annihilate much of human society by the systematic destruction and contamination of world ecosystem biodiversity and productivity?

The Hegemony of Microorganisms

The Earth's biosphere is one huge ecosystem constructed from a subordinate system of interdependent biome ecosystems. The combined effect of rapid growth in world population and the emergence of a global consumer culture that produces massive amounts of chemical, industrial, and biological waste (entropy), results in a natural, organic increase in the contamination, degradation, and disruption of the Earth's biogeochemical cycles. Efficient ecotoxin uptake and transfer occur throughout all trophic levels. The billions of many thousands of varieties of bacteria in every cubic millimeter of organic soil, as well as the wide variety of other microorganisms, such as fungi, molds, and phytoplankton, absorb, metabolize, and recycle thousands of anthropogenic ecotoxins on a molecular level far below current limits of observation and documentation. As anthropogenic activity alters and then destroys ecosystems, resource depletion, the loss of biodiversity, species extinction, and anthropogenic ecotoxin uptake and biomagnification constitute a naturally-occurring, organic process. The increase of entropy in human ecosystems, as industrial and post-industrial societies construct ever more complex and vulnerable social and economic infrastructures, will result in a decline in the world's human population and productivity as an inherent and expected consequence of this process. Human society is a collective intelligence that may never acknowledge the transience of its presence in the context of a biosphere inhabited by a googolplex of microorganisms that have dominated the biosphere for millions of years.

Biocatastrophe as a Plume

Biocatastrophe may be visualized by mapping; color coded schemes of oceanic dead zones or high incidence terrestrial chemical or radioisotope fallout areas documented by modern scientific data collection technologies (e.g. aerial radiological surveillance). Biocatastrophe is characterized by growing plumes of ecotoxins, which impinge on shrinking areas of agricultural productivity and healthy ecosystems with long established biodiversity. Plume movement can be tracked in time; the decline of some banned ecotoxins (e.g. DDT, tetraethyl lead, etc.) in biological monitoring surveys are offset by the long chemical lives of other persistent organic pollutants, e.g. PCBs, and their continuing distribution and biomagnification in pathways to human consumption. Present and future monitoring strategies are and will be characterized by analysis of multiple ecotoxins in media such as water, breast milk, maternal cord blood, and food products as measured by nano-, pico-, and femto-technologies. The synergistic presence of multiple neurotoxins will be the inevitable subject of future plume analysis.

Biocatastrophe Timeframe

Biocatastrophe is a simultaneous field of phenomena acting synergistically, but in different timeframes. Resource destruction occurs at different times in different ecosystems. The historical evolution of biocatastrophe and its relationship to growing human populations and dwindling natural resources suggests that the impact of major components of the phenomena of biocatastrophe will be felt within the next century. The timeframe of the impact of chemical fallout has already started; its impact in terms of the chemical half lives of persistent organic pollutants will be measured in hundreds of years. Unsecured anthropogenic radioactive waste sites, especially in the former Soviet Union, but also in the US, have future impacts that will be measured in thousands or tens of thousands of years. Future economic and infrastructure collapse of the global economy may coincide with rapidly disappearing oceanic fish stocks or unexpected disruptions in the efficiency of genetically engineered industrial agriculture. The timeframe for these components of biocatastrophe is ±50 years. The current global financial crisis may be a temporary recession in the accelerating growth of an unsustainable global economy; it also could be the early stage of infrastructure collapse of a global economy that has already reached its limit of technometabolism. Informed debate about these two alternative scenarios is increasingly unlikely given the rapid marginalization of the "educated elite" who would be the prime movers of any solution to the future challenges of the phenomenology of biocatastrophe.

Joseph Tainter on the Collapse of Complex Societies

"Complex societies historically are vulnerable to collapse, and this fact alone is disturbing to many. Although collapse is an economic adjustment, it can nevertheless be devastating where much of the population does not have the opportunity or the ability to produce primary food resources. Many contemporary societies, particularly those that are highly industrialized, obviously fall into this class. Collapse for such societies would almost certainly entail vast disruptions and overwhelming loss of life, not to mention a significantly lower standard of living for the survivors... only a veneer of complexity lies between us and the primordial chaos, it is thought, the Hobbesian war-of-all against all... it is easy to overemphasize such matters, for only a small part of the population is actively preparing for collapse. On the other hand, no educated person who is aware of historical collapses can escape occasionally wondering about current conditions... Certainly none can argue that industrialism will not *someday* have to deal with resource depletion and its own wastes. The major question is how far off that day is. The whole concern with collapse and self-sufficiency may itself be a significant social indicator, the expectable scanning behavior of a social system under stress, in which there is advantage to seeking lower cost solutions... As in the study of historical collapses, those concerned about current conditions have ignored the principle of marginal returns on investment in complexity." (Tainter 1990, 209-10).

Unexpected Events

The dynamics of biocatastrophe are enhanced by unexpected events, many of which will result from either natural occurrences (hurricanes, earthquakes, volcanic eruptions, pandemics), or human activities such as warfare and terrorist attacks. The worst case scenario for an unexpected event is the possibility of a nuclear attack by one country or a terrorist group on an operating nuclear reactor. Nuclear vaporization of an operational nuclear reactor and its spent fuel pool has the potential to release 50 to 100 times the biologically significant radioactivity released by the Chernobyl accident, resulting in contamination of a large portion of the world's food producing ecosystems. The September 11, 2001 destruction of the World Trade Center is an example of an unexpected event with vast national and global consequences. This event signaled the extension of regional sectarian warfare (jihad) on multiple continents, fueled in part by the growing numbers of individuals disenchanted by or opposed to global consumer society. The inevitable participation of the United States in warfare where there is no possibility of military victory wastes huge amounts of financial resources in an economy already so indebted by the activities of a predatory consumer culture and its shadow banking network that it could be argued that there may be no recovery from the global financial crisis, which began in 2008. The 911 attack, the spread of regional warfare in Asia and Africa, and the ongoing financial panic may be signals that the first stages of biocatastrophe are occurring now rather than several decades or more in the future.

Flashover Events

As a result of the combination of rapid population growth and the concurrent spread of combustible materials characterizing urban growth, selected communities in unique vulnerable geological environments such as the Los Angeles basin, Mexico City, and some large Asian cities, if subjected to an earthquake of more than 7.5 on the Richter scale are at the risk of a flashover event. In the case of the Los Angeles basin, the combination of dry and windy conditions and large supplies of combustible fuel, housing, consumer and electronic goods, and natural foliage, including a web of natural gas pipelines and large numbers of motor vehicles, can result in a flashover event in much or all of the Los Angeles basin in ±30 minutes in a worst case (high Santa Ana wind event) scenario. The advent of modern electronic communication systems marks a new era where regional biocatastrophes of whatever causes (e.g. Hurricane Katrina, the Gulf oil spill) can be viewed and experienced by all the participants in the global consumer culture. Such regional biocatastrophes, it may be argued, prophesize the eventual fate of our ever more technologically sophisticated global consumer culture.

Anthropogenic Ecosystems Infrastructure Collapse

A complex industrial society that is dependent upon uninterrupted growth and the infinite availability of non-renewable resources achieves global bankruptcy in stages that correlate with the increasing unavailability of nonrenewable resources and the proportionate decline of sustainable economic activities based upon renewable energy and natural resources. The recent and ongoing global financial crisis is rooted in the egregious and unfortunate activities of an unregulated free enterprise market economy and its predatory real estate and banking interests and is a prelude to the coming age of anthropogenic ecosystems infrastructure collapse. The collapse of our global industrial market economy is inherent in the short sighted values of an anti-ecological global consumer culture based on the model of the narcissistic, unregulated American Environmental Fascist Free Enterprise System (AEFFES). The concepts of an ever-growing global market economy and biosphere as bank account are the ephemeral illusions of an indulgent, decadent, hedonistic consumer society where the systematic

104

dispersal of ecotoxins is the invisible counterpart to the painfully obvious consequences of a bankrupt global consumer culture. The toxic assets of overvalued real estate, stock market portfolios, and huge corporate and investment debt have at least temporarily torpedoed the rapid growth of global consumer society in developed nations. Massive legacy obligations have now been accumulated by all well-intended western governments. World ecosystem biodiversity and productivity are rapidly declining while glaciers and sea ice are rapidly melting. What is next?

The Irony of Annihilation

Modern society is trapped within a unique Pandora's Box with multiple lids; open just one and you have a disaster of epic proportions in the making. The most visible challenge facing humanity is the global financial crisis involving the accumulation of huge quantities of debt, recently noted by Niall Ferguson in a 2009 MPBN-TV show on the ecology of money as 596 trillion dollars. Other highly visible humanitarian crises include worldwide malnutrition and genocide in African communities. Rapidly spreading unemployment, asset devaluations, and growing food insecurity are the most visible components of the world financial crisis. Also obvious is the long term threat of the lack of financial resources by governmental as well as NGO organizations to fund essential health care needs, public education, and the many upcoming unfunded legacy costs of well-intentioned modern society. Hidden underneath these accelerating challenges to a highly leveraged global consumer society, that soon may sink due to its burden of indebtedness, is the cumulative legacy of decades of the systematic discharge of anthropogenic ecotoxins of every description into a biosphere with vulnerable ecosystems and easily extinguished biodiversity. The American free enterprise system and the global market economy have clearly flourished due to the economic profits and comforts obtained from the many inventions and consumer products of a sophisticated industrial society. The use of pesticides, fire retardants, plasticizers, solvents, personal care products and pharmaceuticals, growth hormones, and other inventions of petrochemical and bioengineering industries are the essential components of our comfortable modern lifestyle. Their production and use have been very profitable for western market economies and the global consumer culture, which is their progeny. The irony of the annihilation of living ecosystems implicit in genobiocide is that the postponed costs of these wonderful products of modern technology, the ecotoxins that are their legacy, now play a major role, despite their relative invisibility, in the phenomenology of biocatastrophe.

The Sinking of the Titanic, 4/15/1912 at 2:21 AM

The fate of our world ecosystem, the biosphere, and the global military/industrial/consumer society that now dominates all natural ecosystems can be

105

equated in a metaphorical sense to the sinking of the Titanic. Its sinking was a startling prophecy about the ego-driven fate of a global consumer culture driven by the sectarian beliefs of an ironically atheistic amalgam of Christian and Environmental Fascists with a fondness for violence and conspicuous consumption. The visible portions of the iceberg, which could easily be seen but not heard in the calm seas of that fateful night, can be equated with the rapidly growing indebtedness of a narcissistic consumer culture that tore apart the essential watertight compartments (world commercial and shadow banking structure) revealing the vulnerability of a supposedly unsinkable ship. Of even more momentous significance is the invisible, submerged portion of the iceberg, which facilitated the sinking, a metaphor for our inexcusable contamination of the world's biosphere with anthropogenic ecotoxins. Approximately 25% of the Titanic crew and passengers survived the sinking; many first class passengers, or what we might call the techno-palisaded elite, did not survive. On the 100th anniversary of the sinking of the Titanic, how far along the path to biocatastrophe will we be and who will be the survivors? When the Titanic hit the iceberg, few on board had any inkling of its final fate. As we hit the iceberg of unsustainable human activities, how long a delay before we realize the fate of a biosphere with limited renewable and nonrenewable resources? Which cultural communities will make it to the lifeboats? What proportion of the techno-palisaded elite and their (our) security apparatus will survive along with the working under-classes to keep the trains (lifeboats) running on time?

The Context of Biocatastrophe

Biocatastrophe occurs in the context of a global economy where demand for essential commodities exceeds the supply. The economic elite, that top 5% of 6.82 billion people (World Population Clock as of May 20, 2010), can still afford the gasoline, aviation fuel, whole foods, fresh water, and private health services that are becoming progressively more expensive for lower income groups, including increasing numbers of citizens formerly living the middle class lifestyle of global consumer culture. In the age of sudden disruptions of financial markets and falling asset values, the upper class of privileged academics and middle and lower echelon managers will gradually become the middle class. Many in the middle class will join an ever vaster lower class that does not have access to, or cannot afford, essential commodities, healthy foods, fresh water, and the health services once available to a much larger percentage of the population. Restricted opportunities and resources, the rapidly rising burden of indebtedness, accelerating commodities costs, and a decline in economic prosperity will increase stress and result in downward social mobility. The growing intensity of invisible contaminant signals, especially the worldwide proliferation of endocrine disrupting chemicals (EDCs), which is the legacy of global consumer culture, will impact public health, increasing the demand for ever more costly medical services, and will eventually result in ever-more limited health care options. All serve to increase the vulnerability of affected populations to one of the most important components of biocatastrophe, its health physics impact. Increasing cancer rates, the proliferation of immunological disorders, the appearance of new strains of antibiotic resistant bacteria (ABRB), increasing stress and uncertainty, and, in children, rapidly rising rates of autism spectrum disorders, obesity, asthma, diabetes, and behavioral dysfunction all characterize the age of biocatastrophe.

The Death of Gaia

The collection of drinking water and natural food is a primordial human activity. The fundamental characteristic of human culture is the creation and use of tools to harvest tangible products from natural resources. The harvesting of natural resources and their conversion to tangible products and artifacts is a fundamental anthropological activity of human culture. Pyrotechnology is also a defining characteristic of anthropological activity; the use of fire to produce glass, pottery, brass, iron, and steel provides a narrative of the birth of the age of chemical fallout. In the modern era of global warfare, now the post-Cold War era of regional warfare and growing social unrest among the disenfranchised, the production of weapons, military equipment, and fuels creates huge inventories of chemical wastes. Global industrial/consumer society also produces its own unique blend of ecotoxins. The evolution of massive industrial, agricultural, and information technology ecosystems and their proliferation of pesticides, genetically modified crops, growth hormones, pharmaceuticals, nanotoxins, and electronic wastes have the ironic potential to undercut the evolution of sustainable agriculture. The multiple sagas of global and regional financial crises are the most visible components of the impingement of human ecosystems on natural ecosystems. The evolution of a uniquely American ethos – the American Environmental Fascist Free Enterprise System – exacerbates the propensity of human civilization to commit collective genobiocide. The visible physical destruction of the world's ecosystem biodiversity and productivity is accelerated by the invisible impact of anthropogenic chemical fallout. The death of Gaia, the tragedy of the World Commons, is the cybernetic collapse of the biotic infrastructure of the world's ecosystems due to the contamination and destruction of the biogeosphere by human activities. The viability of human cultural ecosystems, now dominated by a predatory corporate elite, is further undermined by the worldwide growth of a large angry underclass of citizens living on the periphery of the benefits of flat-world information technology. In America, social and political disenchantment takes the form of a rapidly growing, if disparate, community of "Tea Party" dissidents who make a major contribution to our declining capacity for informed debate, effective bipartisan political action, and our lack of financial resources to combat infrastructure collapse. The phenomenology of biocatastrophe is an ongoing series of interrelated social, political, economic, and environmental events. Its net impact will be the radical alteration – the implosion, as it were – of many of the human ecosystems that now constitute American culture, western market economies, and a rapidly globalizing World Commons.

XIV. The Unfolding Age of Biocatastrophe

Convergence

We are now living in the early stages of biocatastrophe: the synergistic interaction of overpopulation, a non-sustainable global military/industrial/consumer society, high indebtedness and financial markets collapse, a vast and accelerating proliferation and worldwide dissemination of anthropogenic ecotoxins, the inevitability of cataclysmic climate change, multiple emerging and reemerging pathogens, and the certainty of a new world of deliberately or inadvertently modified genetic organisms. Greenhouse gas emissions and cataclysmic climate change crisis are only two slices of the biocatastrophe pie chart. Only the future genomic manipulation of living matter by information age bioengineers may yet rival the impact of the past accomplishments of the inventive chemists who gave birth to our petrochemical industry or the more recent emergence of the financial engineers of our shadow banking network. As we enter the unpredictable early stages of biocatastrophe, the time line for this phenomenon is unknown and possibly unknowable, but as world indebtedness approaches 600 trillion dollars, the social and economic implosions and infrastructure unraveling that will be an integral future component of biocatastrophe may already be at hand. At some point in time in a finite biosphere composed of biogeochemical cycles that unite all ecosystems into one global world ecosystem, we are confronted with the inevitability of the convergence of multiple, destructive, non-sustainable activities of narcissistic pyrotechnic global military/industrial/consumer society: flashpoint.

Flashpoint

Flashpoint is when human civilization, powered or characterized by internal combustion and nuclear waste-producing heat engines, endocrine disrupting chemical fallout, and an out of control shadow banking kleptocracy drives over an ecological cliff. Energy, time, and labor-saving information technologies are the prime movers of now globalized warfare, trading, transportation, consumer product manufacturing systems, for-profit GMO producers, and a predatory multi-national corporate and shadow-banking elite, the most important players in the race to ecological oblivion. Flashpoint occurs when the Earth's biosphere, acting as a cesspool for anthropogenic wastes, has no more capacity for ecotoxin absorption. Human political ecosystems, now dominated by vociferous reactionary minorities and silent finance capitalists and their Tea Party allies have lost their capacity for informed debate, effective bipartisan political action, or ability to generate the financial resources to combat infrastructure collapse. Ecosystem biodiversity and productivity are rapidly declining; food and fresh water supplies are characterized by significant ecotoxin contaminant pulses. The majority of the participants in a now globalized consumer culture are and will be too uninformed to understand the subtleties of the invisible externalities of entropy and too angry to effectively deal with the unfortunate consequences of the demise of the viability of our Round-World Commons.

The Decline and Fall of Industrial Society

The unfortunate imposition of human ecosystems upon natural ecosystems rather than their sustainable incorporation within natural ecosystems is an inherent component of the tragedy of the World Commons. Can anyone escape from that biosphere ballroom World Commons labyrinth and its endless smorgasbord of ecotoxins, greenhouse gases, ABRBs, GMOs, and the implosion of human ecosystems that naturally follow? Will flashpoint be mitigated or enhanced by the ongoing collapse of a global consumer society in which world debt already far exceeds world assets? With the help of America's Tea Party reactionaries or their economically marginalized counterparts in any nation, will the surviving financial elite succeed in hoarding scarce financial resources that could be used to mitigate the consequences of flashpoint? Will interest payments of tens of trillions of dollars per year on world debt, now the lucrative profits of our predatory banking equity, hedge fund, and stock market gambling casinos, consume our finite economic resources? How will our children and grandchildren pay for health care, education, environmental monitoring, employment opportunities based on investments in sustainable economic activities, and the assurance of a minimum of social and psychological security for all? Will we know the moment when we as a species drive over the cliff of disintegrating natural and human ecosystems? Will any survivors cling to the cliff wall on the way down, or have a soft landing at the bottom? What is the fate of our children's children and the generations who may or may not follow us?

Hope (January 20, 2009)

A political change of guard occurred in America at 12:00 on January 20, 2009. While a predatory financial power elite and sectarian ideologues remain entrenched within the labyrinths of American society, a spirit of change, renewal, and hope – a "new era of responsibility" – emerged as a result of the transfer of presidential power to the Obama administration. The rejection of the military, economic, and social policies of the Bush administration by a small majority of American voters raised the hope that an open debate would occur on the subjects of most importance to the future viability of American and world society.

- How to deal with the accumulated indebtedness that is our legacy to our children for generations to come

110

- How to mitigate and, in fact, change our collective historic propensity to consume the finite resources of our biosphere with non-sustainable and often predatory activities and lifestyles

- How to respond to the difficult challenges of cataclysmic climate change and discover and implement technological innovations that will lead to sustainable industrial, agricultural, and commercial activities

- How to acknowledge, document, and mitigate the presence and long term threat of anthropogenic chemical fallout in a biosphere with a vulnerable global atmospheric water cycle

- How to respond to the challenges of the growing threat of antibiotic resistant bacterial strains, newly emerging viruses, growth hormones, pharmaceuticals as ecotoxins, and genetically modified organisms, and mitigate and control their impact and proliferation

- How to deal with growing national and world unemployment, income and health care disparities, food insecurity, and global water stress

- How to confront and solve the numerous scientific, economic, social, and political challenges posed by the looming world environmental catastrophe we call "biocatastrophe."

Can we return to a more open society with less control of national and world dialog by vitriolic sectarian ideologues? Can we mitigate the impact of a predatory shadow banking network that profits from the use of arcane mathematical formulas and super computers to manipulate stocks, bonds, and collateralized debt obligations, systematically swindling governments, investors, and pension and retirement funds, including worker's 401k funds? Is there the chance that in future years we can replace the controversial and contentious phrase "American Environmental Fascist Free Enterprise System" (AEFFES) with the traditional description "American Free Enterprise System" (AFES)? In the future, will we reaffirm a political system that respects the rights of all American and world citizens, protects the environment, restores the "ideals of our forbearers," and acknowledges the difficult challenges posed by the unfolding Tragedy of the World Commons, while maintaining America's traditional values of innovation, entrepreneurship, and justice for all?

November 12, 2009 Update: Economics as Superfreakonomics[3]

Taking the collapse of Lehman Brothers on September 15, 2008 and the worldwide disruption in financial markets that followed as the iconic beginning of the age of biocatastrophe, the following observations can be made about events in the year following the election of the Obama administration pertaining to economic, political, technological, and environmental issues, many of which have obliterated the hopeful optimism accompanying Barak Obama's election.

- The evolution of a world financial crisis and its partial and temporary mitigation by massive government intervention has been characterized by what is now an increasing awareness of the unsustainability of a resource-devouring global consumer society based on the use of debt to stimulate economic growth and nonrenewable fossil fuels to create energy.

- An evaluation of the dynamics of global consumer society and its manufacturing and trading activities, the driving force of globalization, indicates economic growth is continuing in "developing countries" (China, India, South Korea, Indonesia, Brazil, etc.) not burdened with huge governmental, commercial, and personal debts of the U.S. and Western market economies.

- In contrast, in faltering Western market economies consumers and government are having difficulty paying the interest on their debts and may never reestablish their former hegemony in global financial markets.

- Unless U.S. Federal and commercial debts are dramatically reduced, continued use of the dollar as the world's reserve currency will likely end within one or two decades.

- The growing economies of developing nations will furnish less consumer products to financially stressed western market economies, while benefiting from rapid domestic increases in consumption of their own manufactured goods.

- Growing Third World populations in most other nations cannot afford to participate in a fossil-fuel-driven global consumer society. Industrial agricultural will continue to undermine the self-sufficiency and sustainability of their productivity; increasing food and water stress are the legacy of this accelerating marginalization.

- The response of the U. S. and European governments to the unexpected collapse of an overleveraged commercial and investment banking system and its predatory shadow banking network of hedge funds and derivatives marketing has resulted in a massive public bailout of the laissez-faire Reaganomics-based finance capitalism casino.

- There may be insufficient financial resources available within the European banking community to mitigate the unfolding euro currency crisis, which has the potential to trigger a second financial meltdown in American and global financial markets.

- In America, the effort to rescue a faltering economic system approaching a total meltdown has resulted in a rapid increase in national debt. No effective attempt has yet occurred to regulate the shadow banking network that was the root cause of the financial meltdown. The banking regulation reform legislation being negotiated in Congress in the summer of 2010 is unlikely to restrain the predatory speculative inclinations of a now entrenched shadow banking industry.

- Informed awareness and nonpartisan debate about the ongoing unraveling of our debt-ridden post-industrial society are casualties of the increasing marginalization of the "intellectual elite" who are the essential enablers of this debate.

The Politics of Disenchantment

One of the most important developments in the first year of the Obama administration has been the continued growth of political factionalism and the undermining of the consensus needed to facilitate critical social and economic programs, such as health care reform, climate change legislation, and regulation of an out-of-control shadow banking network.

- The economic dislocations and increasing unemployment that characterized the growing subprime real estate fiasco in 2007 and the financial meltdown of 2008 culminated in growing political resentments, which were quickly transferred from the Bush administration to the Obama administration during this first year of frustrated political maneuvering. The political paralysis that has emerged is a natural result of the disenchantment accompanying the increasing economic stresses of 2008-09 and the growing insecurity of American's working and middle classes.

- Informed debate became the casualty of the growing cacophony of FOX News and other electronic media. Reactionary elements in society, such as the Tea Party advocates, joined well funded corporate and lobbying interests to dominate media coverage of both the Obama administration and the essential reforms it was trying to implement as typified by the compromised and delayed health care reform effort. In the wasteland of American television, only a tiny minority of outlets (Public Broadcasting, Frontline, Alternative Radio, and occasional reporting by CNN, NBC, and Sixty Minutes) remain to provide news and debate on the compelling issues of the Age of Biocatastrophe.

113

- American free enterprise enthusiasts and Tea Party advocates now ominously use President Obama and the newly elected Democratic administration as scapegoats for the rapid growth of indebtedness, unemployment, and economic dislocations that have their roots in Reaganomics and the reign of Alan Greenspan.

- The fact that a return to long term growth of the American economy can only be accomplished by the accumulation of additional indebtedness by consumers, investors, commercial credit markets, and the government is a reality yet to be publically acknowledged by the Obama administration.

- The continuing growth of fourth world of disenchanted Muslims and others, many of whom as a matter of faith don't wish to participate in a global consumer culture, has resulted in the further spread of armed dissidents in Pakistan, the Philippines, central Asia, Africa, and elsewhere. As symbolized by 911 and the accelerating diffusion of Jihad, this growing sectarian militancy has the potential to spread violence worldwide as the economic impact of a world financial crisis increases unemployment and social stress in many nations on all continents.

The Rapid Growth of Information Age Technology

An important characteristic of the financial crisis that now accompanies the evolution of biocatastrophe is the rapid growth of fiber optic-based digital technologies that have been the prime movers of globalization. They are also the key component of both sustainable and unsustainable economic growth and productivity of the complex societies of the future. These information technologies have four major downsides:

- The age of information technology is characterized by the accelerated replacement of ever larger segments of managerial, clerical, and manufacturing personnel who are no longer needed to facilitate the productivity of a complex global economy.

- Impacted by unemployment and financial stress, this underemployed component of the middle class will not be able to mitigate the recessionary tendencies of a faltering market economy due to a lack of disposable income and an unwillingness to incur further debt to purchase nonessential consumer goods and services.

- Silicon-chip-dependent information technologies produce huge quantities of environmental contaminants and nanotoxins whose environmental costs and impact will rival that of chemical fallout from traditional fossil fuel and petrochemical-based industrial society.

- Implicit in the rapid expansion of the most productive industries of a rapidly globalizing world economy, the result of the evolution of innovative fiber optic, biopharmaceutical, and nanotechnologies, is the insidious growth of massive income, health care, and educational opportunity disparities. While the economic dislocations in most western economies caused by the 2008 financial crisis continues to grow, 1% of the American population now own almost 50% of its domestic assets. A dwindling percentage of workers (±5%) will benefit from the high incomes derived from modern information technologies and multinational corporations. Most other employed workers will remain on the margins of the financial benefits of the age of information technology, while confronting the growing specter of employment insecurity.

A Changing Environment

The environmental impact of cataclysmic climate change has become the subject of more extensive media coverage. As a result most informed citizens are more aware of the increasing evidence of the acceleration of global warming, sea level rise, polar ice and permafrost melting, and intensified weather events, many due to climate change feedback mechanisms that cannot be controlled by political activities or changes in lifestyle.

- Significant reduction of greenhouse gas emissions from industrial activity are increasingly unlikely in growing Near and Far Eastern economies and in the United States, where political paralysis and the widespread belief in cataclysmic climate change as a myth offset our growing awareness of its probability.

- There is expanding media coverage of the rapidly escalating world water and food crisis as typified by the environmental, social, and political chaos in sub-Saharan Africa and as a result of the earthquake in Haiti. There is more public awareness of links between environmental degradation, food and water shortages, unemployment, and the spread of opposition to Western market economy-derived consumer culture, exemplified by the growing influence of the Al-Qaeda in many countries and the seething social insecurity of hundreds of millions of unemployed workers in many nations.

- The collapse of the 2010 Copenhagen Climate Conference and its attempts to coordinate greenhouse gas emission reductions on a global level was an ominous indicator that control of global warming and the multiplicity of its feedback mechanisms is beyond the political, social, and economic capacity of a rapidly growing global consumer culture to facilitate and expedite.

- Contributing to the collapse of the Copenhagen Climate Conference was the failure of the United States Congress to pass greenhouse gas emissions related cap and trade legislation and thus to provide a firm commitment to the goals of the conference. The paralysis of the political process in the US Congress since well before the Copenhagen Climate Conference makes it unlikely that any effective legislation to reduce greenhouse gas emissions will be passed at any time in the future.

- The American public remains almost completely oblivious to the public safety consequences of the contamination of the atmospheric water cycle and most food webs with chemical fallout, including persistent organic pollutants (POPs), endocrine disrupting chemicals (EDCs), carcinogens, mutagens, and other biologically significant industrial effluents and products.

Irony, Paradox, and Politics

The most challenging problem of the Obama administration in its quest to implement common sense health care and financial regulation reform, greenhouse gas reduction, and other progressive programs is the rapidity of the transference of blame on his administration for the cumulative problems of 30 years of the deregulation of an out of control, debt-incurring, consumer society. While the Clinton administration had balanced the federal budgets, paradoxical but unacknowledged increases in credit market debt and continued deregulation of financial markets in the form of the recision of the Glass-Steigal Act were the invisible prime movers of a consumer society with an addiction to McMansions, SUVs, credit cards, and the frivolous consumption of nonessential goods. After 911, the invasion of Iraq constituted a victory for Al-Qaeda in the sense that it insured the growing influence of a rapidly spreading Jihad. The quagmire in Iraq, and now in Afghanistan, resulted in increasing dissatisfaction with floundering American foreign policy. As the economic dislocations of the subprime mortgage crisis spread across the American landscape in the form of unemployment, empty stores, and abandoned neighborhoods, Obama won the 2008 election on the basis of a surge in Main Street dissatisfaction with the status quo. The unfortunate consequence of the continuing economic dislocation was the rapid transfer of blame for the entrenched problems of a debt ridden consumer society to an administration in political power for only one year. Main Street America, in an age of mass cultural illiteracy, lacks the mental agility and, possibly, the moral integrity to understand the consequences of our half century of overconsumption, heedless reliance on a fossil fuel economy, and increasing indebtedness. The end of the American empire and rapid transfer of economic power to developing nations now utilizing our inventive information technologies to mimic our century of world economic hegemony is a

116

natural consequence of the historic inevitability of the rhythmic cycle of nation state empires (Spanish – Dutch – English – American – Chinese [Phillips 2006]). The irony of Main Street America blaming the Obama administration for the economic problems, which derive from America's unique, but transient, unregulated "free enterprise" market economy, is the unintentional enhancement of the economic dislocations now occurring in the early stages of the age of biocatastrophe.

The Loss of Consensus and the Rise of Disenlightenment

The anger, partisanship, paralysis, fears, media and corporate electronic propaganda, and the reactionary politics that have gradually emerged as the fundamental characteristics of American society in the post-911 era are emblematic of the social and political milieu of the unfolding age of biocatastrophe. The conservative nature of America's open society makes it highly unlikely that effective political action can be taken at any time to address the multiple crises of increasing debt, declining natural and public resources, and diminishing industrial productivity in a rapidly globalizing economy where developing nations will soon surpass the gross productivity of the ancient market economies of Europe and America. The gradual end of the American empire – the post-American world as it were (Zachariah 2009) – means a increasingly large segment of American society can no longer rely on the work ethic, creative innovation, traditional business models, (the family farm, country town general stores, small manufacturing businesses), or reliable employment in any skilled profession. Reactionary anger to the growing lack of opportunity in a cruel new world of predatory multi-national corporations who are also vulnerable to resource depletion, environmental crises, and financial meltdowns is as American as apple pie, even if our pies are now slightly contaminated with invisible EDCs and other ecotoxins. No better example exists of the vulnerability of major corporations to unexpected events, including the end of the age of oil, than the ongoing Gulf oil spill disaster and the huge losses incurred by BP. A sketch of our political milieu – a snapshot of the American voting public one year after the election of the Obama administration – helps explain our collective helplessness to address social, economic, and ecological problems that are decades old. No political party or leader has the power now or in the future to effectively ameliorate the multiple crises that underlie the insipient tsunami of biocatastrophe. The open horizon of America's nearly unlimited landscapes have long allowed the florescence of many conservative, often enterprising and adaptive, political and religious groups who have been liberated from the traditional political and social restrictions of the crowded European landscape or the more totalitarian traditions of developing nations. As the open horizon of the American landscape has become restricted by rapid population growth, the ecological disasters inherent in urban and suburban development, and the recent economic meltdown, the inevitable result is anger and disillusionment with the efficacy of government or its lack thereof. The following sketch helps explain why political solutions to many of the most important contemporary economic, social, and environmental problems are highly unlikely.

2010 Pie Chart of American Politics

Mythical political inclinations	Subgroups	Theoretical percentage of electorate
"Reactionary Republicans"		
Vote no on everything	Traditional conservatives	20%
	Evangelical Christians (mostly white)	20%
	Disillusioned Tea Party enthusiasts (independents)*	20%
	Reactionary media elite**	< 1%
	Wall Street elite***	<1%
	Total percentage of voters	±60%
"Left-leaning Democrats"		
Vote for big government	Traditional working class Democrats^	20%
	Evangelical Christians (mostly black, some Hispanic)	8%
	Marginalized academic, political, and media elite[+]	3%
	Info-tech elite and their well paid employees	5%
	Activist greens[~]	3%
	The rich and the beautiful^^	< 1%
	Total percentage of voters	±40%

*Includes many unemployed white collar and blue collar workers and former undereducated Democrats influenced by fear mongering electronic propaganda.

**Typified by FOX News commentators and other right-wing media pundits

118

***Those rich bankers, etc. loathed by the Tea Party nitwits

^Left-leaning unionists and still employed blue collar and governmental workers

+The detested intellectual elite

~Left-leaning environmentalists who want to undermine our oil economy

^^The super rich, including popular sports and entertainment idols (Bono, Oprah, Springsteen, Jolie), who gave money for the poor and oppressed in third world countries, but not for the dislocated Tea Party enthusiasts in the US

The consequences of America's rapidly changing political landscape is political paralysis in Congress and the consequential inability to address the wide variety of social, economic, and environmental problems that are now undermining the viability of America's free enterprise system as well as causing rapid increases in the cost of health care.

Flat-World Technology, Round-World Politics

As Thomas L. Friedman and many others have noted, digital technology and the "flat-world" of information availability, instant communications, and efficient industrial production has altered the world's economic and political landscapes. Hundreds of millions of world citizens have been lifted out of poverty or the confining restrictions of traditional lifestyles by the new age of information technology. Millions of workers in India and China, formerly earning less than $1,000 per year, have seen their incomes rise by an order of magnitude. They have become the enablers of a global market economy dependent on the digital equipment they now operate. These new information technologies also contribute to the growing control by a small number of multinational corporations and the elite financial market engineers and entrepreneurs of the world's rapidly globalizing economy. This rapid change in the distribution of income and employment opportunities occurs at the expense of the industrial workers who don't share the electronic skills of the younger generation or the white collar workers whose jobs are now obsolete. Flat-world technology, in fact, cuts through a round-world of economic and political communities, which are multidimensional (not flat), complex synergistically interrelated ecosystems, all dependent upon a round-world of natural ecosystems. Electronic distractions join electronic anxieties, diverting tuned in flat-world participants from confronting the multitude of real world issues that lie outside of the flat-world reality of computers, cell phones, iPods, YouTube, Facebook, iTunes, twitter, and the recreational use of electronic technology. Survival in the real world necessitates transcending the flat-world Saturn's ring of information age distractions now surrounding the round-world of earth's political, social, and ecological complexities, and confronting the multiplicity of crises now facing participants in a global economy that is, in fact, not flat. The challenge of the 21st century is how to engender a broad based engagement with that complex labyrinth of economic, social, political, and ecological crises now occurring in our very round biosphere.

Populism in the Age of Super Freakonomics

A fundamental characteristic of politics in the unfolding age of biocatastrophe is the reactionary response of large numbers of people to the attempts of government to mitigate the impact of the unraveling of the infrastructure of American society. The primary thrust of the anger of what is essentially a populist movement is opposition to the institutions of government that should have an important role to play as the finance capitalism – petro-political world of American society disintegrates. Tea Party reactionaries are against everything: abolish the IRS, EPA, banking regulation, gun control, immigration, etc. A recent and continuing lightning rod issue is health care reform, allegedly characterized by too much intrusion of government into our private lives by what is already the bankrupt bureaucracy of Medicare – Medicaid – Social Security. The anti-Wall Street component of this populism, based on valid real life economic dislocations as its prime mover, is anti-progressive; political affiliation, Republican, Democrat, or Independent is irrelevant. What is most important is opposition to regulation, taxes, environmental legislation, and government in general. The irony of the populism of the Tea Party movement and its hope for a return to prosperity and its opposition to the Federal Stimulus Bill is that it is the natural ally of Wall Street and special interests who have an equal dislike of government intrusion and regulation. The reactionary philosophy of populism has close connections with the writings of Ayn Rand, many elements in common with the National Socialism movement in Germany in the 1930s, and is reflected in the vast wasteland of popular books by authors and possible future political candidates such as Sarah Palin and Glen Beck. The Tea Party movement echoes the bankrupt philosophy of the one dimensional Chicago School of Economics, hardly more insightful than what Paul Krugman (2008) notes as the east coast (read: Harvard) school of economics. The chaotic economic milieu of an "unregulated" free enterprise system, the implosion of which is the context of the reactionary simplistic economics of Tea Party fascism, is appropriately referenced by the titles of the bestselling books, *Freakonomics* (Levitt 2005) and *Super Freakonomics* (Levitt 2009). These popular texts trivialize the debate about America's economic crisis and embody the ongoing ritual of aversion to informed debate about the roots of this historic event. No words, however, better summarize our current economic predicament or describe the future economic milieu of an imploding global consumer society than the term "Super Freakonomics".

Media Propaganda and the Manipulation of Public Opinion

January 20, 2010 marked the election of Scott Brown as a Republican senator to occupy the seat formerly held by Edward Kennedy. Initially described as Obama's "Waterloo", this historic event may have signaled the early end of the hope noted above as characterizing the election of the new administration. The anger of an ever-growing reactionary community of unemployed or underemployed or economically challenged citizens from all political parties signals the ominous unraveling of the political consensus needed to insure the survival of an American society in crisis. There is almost now no chance that informed debate, followed by effective political actions on the issues noted in *Hope January 20, 2009* in the previous section, will occur in the context of the growing economic stress experienced by the vast majority of Americans who do not benefit financially from the innovations of the flat-world of information technology. The fundamental reality of American politics and its debt-ridden economy is that the corporate elite, including not only Wall Street but health care and pharmaceutical giants, oil and industrial agricultural behemoths, and other multi-national companies, can, and obviously, have, used modern electronic media to manipulate public opinion about subjects ranging from health care, greenhouse gas/carbon offset, and financial reform legislation to human health issues and consumer product purchases. The widespread, if often subtle, control of electronic media content has the cumulative impact of infusing fear, anxiety, and thus, anger, in a vast segment of America's disenchanted and economically stressed populace. A critical mass of fundamentalists, conservatives, the unemployed, and industrial, white collar, and other workers have coalesced as a powerful "disenlightened" force (Phillips 2006). This social phenomenon can be described as a virtual fiery explosion of discontent in a night sky, the stark reality of the beginnings of the social disintegration characteristic of the early stages of the age of biocatastrophe. Opposed to the moderate reform efforts of the Obama administration, America's corporate plutocracy and health care and sectarian anti-government activists, as well as oil and Wall Street investment banking interests, have been very successful in using subtle and not so subtle media propaganda to exploit the well founded fears of a now nearly disenfranchised segment of American society. It was, and is, in the interest of corporate plutocracy to align this group with the evangelicals and, until now, directionless conservative Republicans forming a formidable, if inarticulate, coalition of reactionary opponents to any and all legislation supported by Democrats. That more lively and less predatory component of "the US debt and credit industrial complex," the techno-elite of Microsoft, Google, Apple, et al., a majority of whom may have supported health care reform, the reduction of greenhouse gas emissions, alternative energy, and financial reforms, have been marginalized by predator corporate America's ironic ability to make angry Tea Party fascists their most powerful ally.

Biocatastrophe Time Line

After six decades of sustained over-consumption of both consumer products and fossil fuels, a sketch of the time line of the evolving phenomenon of biocatastrophe can now be proposed. Two assumptions underlie this sketch 1) that biocatastrophe is a worldwide tragedy of the commons that is occurring, and will occur, over a period of decades and centuries and 2) that the advent of the age of biocatastrophe has become abundantly clear with the financial crisis of September 15, 2008 (the collapse of Lehman Brothers) and the near "total financial meltdown" (a term that remains unexplained by knowledgeable economists) of the world's banking system. This sketch is made in the context of a flattened world characterized by efficient labor and timesaving information technologies (Friedman 2006), and also by an infinitely more complex *round-world* of economic, political, social, and ecological infrastructure systems and problems.

- The economic dislocations of biocatastrophe are its most visible and painful components. All literate world citizens are becoming acutely aware of the vulnerability of their national economies after the world financial crisis began in 2008; this financial upheaval is the most obvious indicator of the beginning of the age of biocatastrophe. Continued economic dislocations and the resulting anger and social and political turmoil will continue for decades, if not centuries, until the post-apocalypse (read Post Armageddon) World Commons (human ecosystems within natural ecosystems) achieves a sustainable balance.

- The ecological components of biocatastrophe are also ongoing and can be divided into the visible and invisible components of a rapidly deteriorating biosphere, with loss of ecosystem diversity and productivity much more visible than the hidden impact of chemical fallout. The recent Gulf of Mexico oil spill is a dramatic, highly visible example of our dependence on the non-sustainable use of nonrenewable fossil fuels and their extraordinary external costs. All the ecological components of biocatastrophe have an economic impact that is not immediately obvious; angry Tea Party activists in America have no interest in, awareness of, or concern about the complex labyrinth of political, social, and environmental issues that are the principal causes of their economic disempowerment. With respect to the Gulf oil disaster, they have been strangely silent.

- The vast health physics impact of invisible endocrine disrupting chemicals (EDCs), persistent bioaccumulative toxic chemical (PBTs), methylmercury, and other anthropogenic ecotoxins are also ongoing and will greatly expand as developing nations, such as China, India, and others, adapt American petrochemical-based consumer product manufacturing systems. The multiple health physics crises which result, many of which are already underway, will continue as long as a globalized industrial society uses petrochemicals and nanotechnology to produce ecotoxins. Ever increasing health care costs are an implicit component of this historical event, i.e. the global spread of modern consumer society.

- The evolution of cataclysmic climate change (CCC) due to greenhouse gas emissions and their many feedback mechanisms has a long delayed economic, social, and ecological impact measured in decades, if not centuries. Worldwide skepticism of its significance is due, in part, to the immediate effect of the economic dislocations caused by the world financial crisis. Developing nations now less impacted by the world financial crisis have a natural vested interest in greenhouse gas-producing growth of their own gross domestic productivity (GDP). Unless suffocated by air pollution or sickened by contaminated water, concern about ecological issues in developing countries in the age of growing global consumer culture, as in America, will be minimal.

- The maximum impact of our huge credit market and legacy debts has yet to occur in a society still dependent on a fossil fuel economy. A bipartisan consensus about the threat posed by our national indebtedness has already emerged as a component of the economic anxieties of the early stages of the age of biocatastrophe.

- Cultural awareness of the links between overconsumption, petrochemical dependence, chemical fallout, increasing health care costs, rising personal debt, and the economic dislocations and anxieties now buffeting most of American society has been submerged by the ongoing demonization of our political process, its leaders, and the functions of government.

- As with the ecological impact of a global consumer society, the maximum impact of the consequences of the use of debt as the prime mover of global consumer society will occur over a period of decades. A late middle age florescence of global consumer culture, based primarily on the temporary hegemony of China as the world's most rapidly growing economy, but also influenced by a temporary spurt of economic growth by Brazil and other South American nations, may or may not postpone the consequences of debt as the prime mover of the decline of a now globalized "free" market economy.

Rituals of Aversion and Denial

A disappointment of the Obama administration is its understandable reluctance to articulate the controversial relationship between the unsustainable elements of our credit market debt driven economy and the ongoing mega-destruction of the productivity and biodiversity of the World Commons as one giant interrelated ecosystem. Facilitating this collapse are an entrenched shadow banking network in need of regulation, our continuing dependence on the use of fossil fuels, and the likelihood of accelerating financial crises in a finite World Commons where the continued growth of global consumer society is not sustainable. Adding to this challenge is a splintering social and political consensus about how to reconcile increasing social needs and declining assets, and the necessity of finding a coherent response to multiplying and increasing expensive public health threats, including the growing impact of endocrine disrupting chemicals, the ever present possibility of pandemics, a worldwide food and water crisis, and the growing cost of health care. These and other crises occur in the context of a biosphere with a contaminated atmospheric water cycle, dwindling productivity, finite assets, and a consumer society that has run out of money. The essential elements in the evolution of biocatastrophe as the inevitable climax of a resource-devouring global consumer culture are increasingly obvious in the year since the election of the Obama administration, yet also continue to be the subject of elaborate social and political rituals of aversion and denial. The five-century-long reign of a theoretically ever-growing Western market economy is now entering its final period of gradual, then rapid, decline, in a world of multiplying, accelerating ecological, political, and economic inconvenient truths.

Hot, Crowded, and Polluted

Thomas L. Friedman's 2008 publication *Hot, Flat, and Crowded* is a most important expression of a central concern of America's increasingly marginalized educated and academic elite. Friedman asserts that Americans must lead the way in a green revolution that then will be followed by China and other developing nations if the rapid increase in global warming is not to become cataclysmic climate change crisis with a worldwide impact. Friedman's primary focus, as is that of Al Gore, is on the greenhouse gas emissions resulting from what Phillips (2006) and many others note as our dependence on a "military petroleum industrial complex," which now threatens the viability of American society. Not mentioned in Friedman's excellent survey of solar and renewable energy industries, experts, and information sources or in his index is the phenomena of chemical fallout, including POPs, perchlorates, consumer product ecotoxins such as endocrine disrupting PBDEs and phthalates, the proliferation of methylmercury in the food web, the rapidly evolving world water crisis, or the impact of human activities on ecosystem biodiversity and productivity. Words such as biotechnology, electronic wastes, autism spectrum disorders, biomagnification, maternal cord blood, synthetic chemicals, or atmospheric water cycle fail to make an appearance in his text. Friedman's emphasis on hot and crowded limits informed debate on an environmental crisis that is, in fact, not flat and linear but a complex series of interrelated ecological crises – the tragedy of a very Round-World Commons. The economic meltdown of September 2008 followed the publication of Friedman's book. His later exploratory essay in *The New York Times* "The Inflection is Near" (Friedman 2009) illustrates an incipient awareness of the linkage of our energy and global climate crisis with our shadow banking culture of existentially criminal, but currently legal, transfer of wealth from workers and investors to a few mathematical wizards and their corporate handlers, armed with expertise in the electronic manipulation of the "free" market economic infrastructure. A more comprehensive survey of the rapidly unfolding saga of the declining viability of a World Commons on which humanity depends for its survival needs a longer title "Hot, Crowded, Polluted, Infected, Indebted, Unemployed, Uninformed: The Unfolding Age of Tea Party Fascists."

The End of Optimism

A series of historically important but ominous events have occurred during the first years of the Obama administration that undercut the possibility of effectively dealing with the long term impact of cataclysmic climate change due to greenhouse gas emissions and their multiple feedback mechanisms. As noted above, Thomas L. Friedman (2008) articulated a series of optimistic goals pertaining to the greening of America, which would then be followed by China, if we would just lead the way. Even more optimistic were the assertions in Al Gore's *Our Choice, A Plan to Solve the Climate Crisis* (2009) that modern technological solutions are available to solve the crisis; what is needed is the political will to implement these technological innovations. *Our Choice*, written in the months before the Copenhagen Climate Conference on climate change (December 2009) predicted an optimistic outcome of the conference whereby the participants from developed nations, developing nations, and undeveloped nations would join in a concerted action to solve the climate crisis. The conference ended in contentious discord with no significant climate-altering agreements. The Gore text also projects a similar optimism about the passage of carbon dioxide limiting cap and trade legislation, which was passed by the House of Representatives only to languish in a dysfunctional and highly partisan Senate. In light of the growing power and influence of the disparate mix of Tea Party reactionaries from many diverse segments of American society, and the ominous growth in the influence and power of FOX News as a media propaganda outlet for the most reactionary elements in American society, no such legislative reforms are now likely. The recent Supreme Court decision to lift any restrictions on corporate special interest advertising during elections (January 22, 2010) is another nail in the coffin of an optimistic America's ability to mitigate climate change and environmental degradation in the age of biocatastrophe. Sectarian political groups can now join multinational corporate entities to formulate electronic media propaganda that can effectively undermine most future political reform efforts.

The Collapse of Finance Capitalism

America has been blessed with a dynamic, creative business and scientific community, which invented and implemented the fundamental innovations of the age of information technology. These advances, especially the evolution of analog, analog-digital, and digital computers, and their prime mover, modern fiber optics technology, constitute the most productive element in the American economy in an age of declining industrial capacity and rapid globalization. Substantial advances in biotechnologies and biopharmaceuticals are a major manifestation of this creative element in an American economy that still has the highest gross domestic productivity of any nation in the world (25% of world GDP). Unfortunately, by 2010, financial services, which create profit by shuffling, if not manipulating, paper in the form of stocks, bonds, mortgages, CDOs, CDSs, equity funds, credit cards, now constitute well over 20% of American GDP. As the market economy debts of a free enterprise system in crisis rapidly accumulate, the creative component of the American economy – its information technologies – produce an increasingly smaller percentage of GDP, in contrast to the rapidly growing, entrenched, often predatory – and still unregulated – finance capitalism component of the economy. The florescence of this tiny component of American society, essentially a ponzi scheme invented and executed by Harvard-trained "quants," signals the early stages of the collapse of the viability of finance capitalism. In this context, "free enterprise" means the broad social acceptance of the unfettered predatory activities of a plutocracy too ignorant and too greedy to support the American workers and their families who are the prime movers of an economy finance capitalists derive their income from.

Descent into Chaos

American society is now unraveling. The creative innovations of American techno-elite and their invention of the age of information technology are an insufficient force to maintain the United States as the world's leading economy. The rapid adaptation by the highly motivated students of developing nations of our advanced scientific, mathematical, and technological information culture, often learned at United States universities and colleges, and the equally rapid relative decline of the productivity of America's public education systems, are the key elements in our ongoing economic deterioration. The most important component of this descent into chaos is the rapid growth of the cultural phenomenon of a divisive, highly sectarian, reactionary social and political element of American society. Highly decentralized and disorganized, this new political force consists of floundering conservative and moderate Republicans of all persuasions, bickering Tea Party advocates with many disparate interests, a wide variety of evangelical and Christian fundamentalist true believers, and a rapidly growing underclass of angry unemployed blue collar workers, displaced white collar workers, and bankrupt small business owners. Their few shared beliefs include their dislike of the plutocracy – those 10,000 members of the corporate and Wall Street elite – who have effectively bankrupted their lives, and their hatred of government as a threat

126

to their liberties. In this context, the key to the descent into chaos is the declining influence of the traditional American values of informed consent and debate, bipartisan political action on the cultural issues of the time, or support of many of the government's essential functions, especially public safety as expressed in public health care reforms, regulation of the out of control shadow banking network, biomonitoring, and mitigating the ecological and economic consequences of the end of the age of oil. Main Street America inadvertently and ironically supports both the status quo and the accelerating decline of the viability of American society with their reactionary and uniformed anger. This entrenched conservatism, the legacy of Reaganomics, in turn has the potential to end the preeminence of a now debt-ridden market economy where the innovations of the American techno-elite are subsumed by the ideological beliefs that are driving America into its ongoing descent into chaos. The undermining of the basis of the viability of the American economy is well underway. The anger at, and marginalization of, America's educated classes and much of its progressive media elite, is a key step in this descent. The penultimate cause of the descent into chaos is the political sabotaging of the functions of Congress, in particular, and government, in general, by the Tsunami of reactionary factionalism, which prevents and effective and informed response to the challenges of our time. The political chaos characterizing the unfolding age of biocatastrophe is a consequence of our economic decline and, in turn, is a predictor of the inevitable but invisible ecological consequences of 5,000 years of imposing human ecosystems on natural ecosystems. The synergistic interrelationship of these historic events include the ongoing implosion of western market economies that have run out of money, the rapid decline in nonrenewable fossil fuel and other essential natural resources, and the loss ecosystem biodiversity and productivity on which human society depends. The urgent need for thoughtful educated analysis of the complicated scientific issues that help us understand why our World Commons is in crisis, join with the unpleasant political and social realities of modern societies under extreme duress. The result is our biosphere ballroom labyrinth of enigmas and dilemmas, the complexity of which is far beyond our current capacity for informed debate. Among the most important quandaries of the legacy of the reign of petrochemical man is our inability to mitigate its ecotoxic impact due to increasingly scarce public and private financial resources.

Discourse on Armageddon and the Day of Judgment

The suggested title of an upcoming publication on the unfolding age of biocatastrophe, with apologies to Thomas L. Friedman:

Hot, Round, and Crowded

Also Polluted, Infected, Genetically Modified, Hormonally Disrupted, Incontinent, Feminized, Unemployed, Uninsured, Uninformed, Manipulated, Marginalized, Indebted, Disenfranchised, Powerless – and Angry

Subtitle

The Tale of the Unfortunate Decline of Western Finance Capitalism Market Economies in the Decimated Round-World Commons of our Giant Integrated Ecosystem

Haiti, Hope, and the Inherent Spirit of Generosity

The impact of a relatively small 7.0 earthquake in Haiti, which resulted in excess of 200,000 deaths, illustrates the vulnerability of rapidly growing populations to biocatastrophe on a local and regional scale, as did the Katrina disaster. The Haiti earthquake also illustrates several phenomena of critical importance in the unfolding age of biocatastrophe. One is our collective capacity in the post-Katrina world to effectively mobilize public resources, including first responders, such as American and foreign rescue teams, and emergency medical personnel from many countries to respond to the devastation caused by a natural event in a densely populated area. A second lesson is the inherent generosity of individuals of all political and religious persuasions and socioeconomic lifestyles who mobilized to raise money and otherwise provide support for the vast impoverished population of Port au Prince and vicinity. In the coming age of increasingly severe geophysical, climatological, and ecological disasters, one can hope that the generosity and support directed towards Haitians in their time of need will characterize American and other nation's response to future crises. Unfortunately, in a world of rapidly diminishing public, private, and NGO resources and multiplying social, environmental, and climatological crises, this may not be possible. In a more generic political context, the urgent necessity of the immediate future is the challenge of the nonpartisan mobilization of the resources of the American people to solve the immediate social and economic problems caused by income and health care availability disparities and the ongoing demise of the economic viability of the American economy. It is difficult to harness the inherent power of the generosity of all socioeconomic and political classes to solve specific social problems when, in a world of increasing economic dislocation, the urgent necessity of consensus is sabotaged by the narcissism of bipartisan demonization.

128

Survival in the 22nd Century

In the 22nd century, the globalized communications inherent in the "flat-world information technology" revolution will be the key element in the survival, if not the florescence, of sustainable industrial and eco-agricultural activities. The prolonged economic and ecological Armageddon of the 21st century will have resulted in a painful series of judgments on a narcissistic global consumer society rooted in petro-politics, the worship of the internal combustion engine, the overconsumption of scarce natural resources, and the legacy of the subprime intellectual assets of a reactionary sectarian mass culture. Survival in the 22nd century will be contingent on the widespread use of renewable resources (there will be no other choice), a minimal toxic footprint left by human activities, the mitigation of the electronic spread of disinformation by special interest power elites, and a cultivation of the inherent human qualities of civilized humanity. Survival of sustainable components of modern industrial society in a world of finite nonrenewable energy and natural resources will be contingent upon the innovative nonpartisan implementation of three of the most important characteristics of western civilization:

- Creativity: the clever adaptation of new strategies utilizing existing resources

- Ingenuity: the invention of new ecologically sustainable technologies

- Compassion: the humanitarian sharing of the scarce resources of the World Commons in comparison to their exploitation as private property in the 20th and 21st centuries

An irony of survival in the 22nd century is that sustainable economies and viable sociopolitical entities may be expedited, not by the dissonance plagued governmental institutions typified by the American Congress, but by innovative sustainable private and corporate enterprise, often on an individual level, which can circumvent dysfunctional or ossified public institutions and the anger of mass cultural factionalism. In this context, resilience, the key to the survival of human civilization in the 22nd century, will be a function of community values, sustainable social and economic activities, and the mutual support systems they engender in a biosphere of shrinking nonrenewable resources, continuing economic dislocations, social unrest, and diminished public resources. The significance of the ongoing contamination of the atmospheric water cycle, and thus maternal cord blood and breast milk of mothers in all nations, with anthropogenic chemical fallout, continues to be an invisible public safety crisis. In the context of the implosion of the viability of the global consumer society, this key component of the phenomenology of biocatastrophe is not yet a topic of public debate.

Postscript – The Election of November 2, 2010

The election of November 2, 2010 marked the probable end of the Obama administration's ability to implement further creative legislative responses to the now accelerating dissent into chaos of the American cultural milieu. Given the obstructionism and hostility of the Republican minority in Congress, now an overwhelming majority in the House, the Obama administration's accomplishments in its first two years are remarkable. The passage of the Recovery Act after just one month in office played a major role in mitigating the consequences of the financial crisis of 2008-9. The health care reform legislation, despite flaws such as the 1099 reporting requirement, is an historic milestone. The financial reform legislation is at least a modest attempt to control an out-of-control shadow banking kleptocracy. The convening of the National Commission on Fiscal Responsibility and Reform is a symbol of the Obama administration's goal of long term fiscal sustainability. In combination with the election debacle of November 2, 2010, the spiraling gyre of continued economic dislocations, rapidly widening income disparities, the proliferation of vitriolic campaign rhetoric, and the perfection of the use of electronic media to spread propaganda and influence public opinion, signal the premature end to the hopeful optimism of the Obama era. These developments help explain the collapse of the ability of Congress to implement nonpartisan legislation essential to mitigating the economic and environmental crises of increasing needs and declining resources. The potentially suicidal failure to let the Bush tax cuts expire, the likely inability of Congress to increase Social Security and Medicare payroll deductions, the obvious impossibility of a value added tax (VAT) to offset the impact of the underground economy, and the failure to tax the huge incomes of the super rich (see below) are all symptoms of the accelerating implosion of the economic viability of the American free enterprise system for all but a small percentage of the highest income earners. The developing nations of China, India, Brazil, Indonesia, and South Korea will have two to three decades to adopt resource-devouring consumer habits of the American "free enterprise" system of buy-now pay-later before the chickens of peak oil, peak soil, peak oceanic fisheries and agricultural production, peak commodities production, and peak world indebtedness come home to roost. The limiting factor of potable water will be the ultimate arbitrator of world population and economic growth. The battle to establish sustainable human communities based on sustainable economies has just begun.

The American Tea Party Taliban

The rise of an American Tea Party Taliban is characterized by resentment, anger, primitive nativism, empty rhetoric, and a reactionary political "smash the lifeboats" philosophy. The emotionalism of the American Tea Party Taliban closely parallels the angry resentments that are responsible for the worldwide spread of the Muslim Taliban.

The absence of practical suggestions by Tea Party and Republican politicians about how to address the accelerating decline of the power and influence of American society in a rapidly changing globalized economy is an ominous portent of our future inability to meet the challenges of a biosphere in crisis. The irony of the rise of the American Tea Party Taliban is that its core is only a tiny minority of activist voters and politicians (the Sarah Palin – Jim DeMint – Ron Paul crowd), but with the help of Fox News, The Wall Street Journal, and sophisticated electronic propaganda spread by highly paid special interest groups, the majority of American voters have been drawn into the politics of an angry anti-government agenda. This conservative opposition to the Obama administration has been unable to articulate practical economic, political, and social solutions to escalating health care and educational needs, the maintenance of existing legacy obligations (Social Security, Medicare, and Medicaid funding,) or the challenges of Round-World resource limitations and cataclysmic climate change. Tea Party activists and many conservative voters are being exploited by special interest power elites ranging from commercialized health care vendors to agricultural interests trying to protect their costly government subsidies. The increasing control of corporate and shadow banking interests over much of the production and distribution of energy and consumer products is mirrored in their ability to exploit and maximize the benefits of an alliance with Tea Party reactionaries whose most creative suggestions include doing away with the Environmental Protection Agency (EPA), IRS, Department of Education, Public Broadcast funding, and the National Endowment for the Arts. A recent post-election *Wall Street Journal* op-ed articulated the escalating disenchantment of a growing percentage of American citizens with the "well paid elites" of the "two left coasts," including "well-off knowledge professionals," the last vestiges of opposition to the "reform [of] entitlements, taxes, and public spending." On environmental issues, ranging from cataclysmic climate change to chemical fallout, *The Wall Street Journal* reiterated the post-election Tea Party Taliban agenda succinctly: "The expansive pieties of our reigning civic religion, environmentalism… nudges us to conform to 'pro-social' behavior" wasting time on "environmental issues like cap and tax as enlightened social statements." (October 31, 2010, A18). In the age of the Tea Party Taliban and the sinking of a titanic global consumer society, the reigning philosophy is now every person for themselves and to hell with lifeboats.

The Economic Challenges of the Age of Biocatastrophe

Among all the slices of the biocatastrophe pie – cataclysmic climate change, chemical fallout, declining natural resources availability, the collapse of ecosystem diversity and productivity, the world water crisis, proliferating antibiotic resistant bacteria – one piece of the pie has immediate inescapable consequences: what happens when a highly indebted global market economy runs out of money. The rise of the American Tea Party

Taliban is an economic based phenomenon, as were the rise of National Socialism in Germany in the 1930s and the global spread of the anti-capitalism Taliban in the early 21st century. The resentments and anger generated by the uncontrollable technology-fueled leveling of world incomes – Chinese workers make more money while Americans make less – is not a reversible phenomenon that can be addressed by legislative actions of federal, state, and local governments. The empty rhetoric of "anti-big government" or "Obamacare" obscures the consequences of four decades of debt-creating Reaganomics and the realities of a rapidly growing flat-world digital and Clintonomic world trading consumer society. Mitigation of the rapidly escalating economic and environmental consequences of a self indulgent narcissistic consumer society, at least with respect to the now highly indebted national and state governments, necessitates innovative long range nonpartisan political and economic policies. Implementations of creative progressive responses to the cruel realities of Round-World biocatastrophe are now highly unlikely in the context of a paralyzed Congress and a demonized government. Among obvious practical pocketbook responses to the escalating economic and social stresses of our current predicament are:

- A value added tax on all consumer goods except basic food commodities. Tax reimbursements would be given to low income individuals for purchases of food, fuel, and medications. One half of these collections should be returned to state governments as ironically advocated by Tea Party advocate John Rigazio in an advertisement in *The New York Times Book Review*, October 31, 2010, for *America is Now a Socialistic Country*.
- The VAT should be significantly higher for luxury goods not normally needed or consumed by middle class Americans.
- The Bush tax cuts must expire. Even Allen Greenspan recently warned of the dire consequences of not doing so.
- Due to increasing income disparities, as typified by the rapidly rising compensation of American corporate executives, Wall Street shadow bankers, and hedge and equity fund managers, the following tax increases should be implemented:
  - Fifty percent tax on incomes above two million dollars
  - Seventy five percent tax on incomes above ten million dollars
  - Ninety percent tax on incomes above fifty million dollars
- Social Security deductions should be extended to cover all salaries and bonuses above the current cap.
- Small increases in deductions for Social Security and Medicare should be implemented for workers earning more than $25,000 a year with larger amounts deducted for those with salaries above $100,000 a year.

132

- Wealthy individuals with post tax income above a half a million dollars should not receive Social Security payments.
- As recommended by the Debt Reform Commission on November 10, 2010, an increase of the federal tax on gasoline would be a practical way to address the deficit, which will otherwise greatly enhance the social and economic impact of the irreversible consequences of biocatastrophe as an historical event.

Unless these relatively modest economic reforms can be made, the critical issues of lowering national indebtedness and funding escalating legacy costs cannot be addressed. These suggestions stand in contrast to an advertisement in the *The New York Times Book Review* for "*How Republicans can Legally Pay No Taxes, Change America, and Save the World*" (Voltaire 2010). This advertisement reiterates the dominant theme of the Tea Party Taliban. "An already too large and overly intrusive government... risks your future financial security, personal liberty, and in some cases, your life." (*The New York Times Book Review* October 31, 2010, 21). That we can "save the world" and "pay no taxes" is the core ideology of the Tea Party Taliban – but save whose world in a biosphere in crisis with seven billion residents?

If American social and economic institutions are going to adapt to the increasing stresses and challenges of the age of biocatastrophe, essential government functions and entitlements must be maintained. The next financial crisis will threaten the viability of all state, federal, and corporate pension funds, health care programs, and entitlement programs. It will also undermine our ability to fund our already declining educational institutions and to respond to escalating environmental crises such as cataclysmic climate change and the invisible spread of environmental chemicals in pathways to human consumption. The failure to address the rapidly expanding disparities in income, health care availability, and quality of education ensures that the impact of the unfolding saga of biocatastrophe will occur more quickly and with more dire economic and social consequences than would be the case without the rise of a powerful American Tea Party Taliban whose incoherent agenda inadvertently undermines the viability of American society to "establish justice, ensure domestic tranquility, provide for the common defense, [including defense against cataclysmic climate change, antibiotic resistant bacteria, and hormone disrupting chemicals] promote the general welfare, and secure the blessings of liberty." (U.S. Constitution). The fervent religious beliefs and reactionary social values of the Muslim Taliban are unacknowledged role models for the patriotic emotionalism and libertarian self-indulgences of a live free or die American Tea Party Taliban. As the American consumer society fractures and implodes, the emerging reactionary extremism and militarism of the American Tea Party Taliban may be the wave of the future.

November 2013 Updates

A number of noteworthy developments pertaining to issues discussed in this text warrant comment.

Hydrofracking and the Continuation of the Age of Oil

The rapidly expanding technology of hydraulic fracturing (fracking) to recover natural gas (methane) from shale deposits thousands or tens of thousands of feet underground has radically changed expectations for an energy crisis due to diminishing oil supplies. Massive fracking operations in the US have lead to rapidly dropping natural gas prices and have kept oil prices from rising. Numerous inefficient coal fired electricity generating facilities have closed in the US. In contrast, in Western Europe, with many more restrictions on hydrofracking operations and dependence on high priced Russian natural gas, gas prices are much higher than in the US. Ironically Europe is increasingly dependent on coal fired energy plants, including new facilities to be built in the future, which will often utilize coal mined in the US. Europe is also still dependent on aging nuclear power plants, as is the US, all of which are essentially accidents waiting to happen. The exception to this is the more safely designed French reactors. The irony of expensive natural gas in Europe contrasts with the low prices of American natural gas and the intensive development of Canadian oil sand. Use of the proposed Keystone pipeline is almost irrelevant as rail capacity to transport tar sands to be exported to China and India are being rapidly expanded. These developments serve to inhibit the development of alternative renewable energy sources to mitigate cataclysmic climate change. The concept of the end of the age of oil was a mirage.

Fine Particle (PM 2.5) Pollutants

The growing specter of air pollution in China and elsewhere due to the rapidly expanding use of coal for energy and automobiles for transportation is an ongoing world ecological crisis. While fine particle pollution (particulates less than 2.5 micrometers) derived from anthropogenic activities has been widely abated in the US, the prospective increased use of coal in China, India, and Europe promises an increase in biologically significant air pollution. The contamination in China is clearly out of control; widespread media coverage of nearly invisible smog-enveloped Chinese cities is dramatic evidence of this continuing threat. The particulate aerosols comprising this highly visible smog are especially dangerous as carriers by absorption of chemical ecotoxins of every description. Many of the nanopollutants in these blankets of smog are, as yet, undocumented. The constituents of fine particle pollution are well known (sulfur dioxide, nitrogen oxide, carbon monoxide, black carbon), the pathways, and distribution of these particulate nanopollutants and the chemical ecotoxin they adsorb remain unknown. Their health physics impact includes lung cancer, respiratory diseases, heart attacks, birth defects, and asthma. Traffic exhaust and emissions from coal fired plants and cooking stoves constitute one source of fine particulate contaminates, but in the context of the rapidly growing economies of China and India, fine particle contamination has multiple sources and combinations, many not yet documented. The most recent EPA air quality standard for annual deposition is 12 micrograms per cubic meter ($\mu g/m^3$). Its 24 hour fine particle standard is 35 $\mu g/m^3$. In contrast, recent levels of contamination have approached 1000 $\mu g/m^3$ in Beijing and other Chinese cities. While smog and sulfur dioxide pollution has been significantly abated in the US, the rapidly growing economies of other nations assure that fine particle pollutants will be a major future environmental issue.

Microplastic Nanopollutants

Our ever expanding inventory of round world ecological crises, easily evaded by our flat world addictions, preoccupations, and electronic recreational activities, includes other looming but nearly invisible threats. The oncoming age of nanotechnology, a component of our flourishing age of information technology, offers a wide array of nanopollutants, many of which have not yet been documented. These will join the many effluents of our industrial landscapes including Silicon Valley's trichloroethane (TCA) and trichloroethylene (TCE), to enhance the impact of greenhouse gas emissions, to pollute the water cycle, and to contaminate the food chains of our round world biosphere.

Microplastic nanopollutants, most of which are composed of polyethylene or polypropylene polymers, are a rapidly growing component of all marine ecosystems. Many originate from the decomposition of plastics in environments such as the great Pacific garbage patch, but also from polyester and acrylic fabrics, which release fibers during washing. Such plastic nanopollutants, along with pharmaceuticals and ecotoxic chemicals of every description, are now ubiquitous in sewage and cannot be removed by our deteriorating sewage treatment facilities. The glaring need to upgrade such public infrastructure contrasts with the congressional inability to appropriate funding to address this and many other public safety issues.

The End of the Age of Antibiotics

Biocatastrophe occurs when a lot of bad things happen at once. The inevitability of cataclysmic climate change, given our rapidly growing greenhouse gas emissions, is a historical fact and the context for a wide variety of ongoing ecological crises. The most important evolving crisis is summarized in the threat report issued by the CDC entitled *Antibiotic Resistance Threats in the United States, 2013*. The report covers 18 rapidly spreading drug resistant microorganisms. It begins with the three most urgent:

- *Clostridium difficile* (CD or C. diff)
- Carbapenem-resistant enterobacteriaceae (CRE)
  "Up to half of all infections caused by CRE result in death." (CDC 2013, 53)
- *Drug-resistant Gonorrhea*
  The CDC report notes that out of 820,000 cases of gonorrhea, 246,000 exhibited "resistance to any antibiotic".

The CDC report continues with comprehensive descriptions of 15 additional rapidly spreading antibiotic resistant infections. A major factor in the rapid spread of antibiotic resistance is the evolution of gram-negative bacteria, such as NDM-1 (New Delhi metallo-beta-lactamase-1), first detected in Sweden in a visitor from India in 2008. This enzyme is easily spread from bacteria to bacteria and is a major source of antibiotic resistance. As one CDC official noted in an October 22, 2013 PBS *Frontline* production (Hunting the Nightmare Bacteria), drug-resistant superbugs are a signal of the end of the age of antibiotics. The link between the rapid spread of antibiotic resistant bacteria and the widespread use of antibiotics in America's and the world's caged animal farm operations (CAFO) and the generic overuse of antibiotics to control disease continues to be a major factor in the rapid spread of antibiotic resistance.

Biocatastrophe: A Simultaneous Field of Ongoing Crises

These and many other ongoing ecological crises continue to unfold in the context of growing factionalism and congressional paralysis in Washington, an Arab Spring that has turned into an Arab winter, the ongoing humanitarian disaster in Syria, growing world income inequality, and the continued vulnerability of a world banking system based on debt as an asset. A glimmer of hope remains that the progressive implementation of the Affordable Care Act can yet occur despite all the electronic glitches characterizing its introduction. All these and many other political, social, economic, and humanitarian crises continue to occur in the context of growing evidence of cataclysmic climate change. Hurricane Sandy, Typhoon Haiyan, and shrinking ice caps are its most recent and obvious manifestations. Also lurking as a potential trigger for social and economic disruptions of massive proportions is the threat of naturally occurring solar storms and their impact on our now highly vulnerable electrical transmission systems. For a summary of Homeland Security, NRC, NOAA, Royal Academy of Engineering, and other reports on the possible consequences of a severe solar storm, go to www.davistownmuseum.org/solarflares.html. The rapid increase in world population has resulted in the increasing vulnerability of human populations in urban and coastal environments to naturally occurring events, such as hurricanes, earthquakes, tsunamis, and solar storms. More extremes in weather cycles, including drought and an increase in heavy rainfall events, are also a threat to human populations in all locations. Ongoing anthropogenic disasters include chemical fallout, constituted by the growing industrial production of over 10,000 ecotoxins, and greenhouse gas and ozone layer depleting chemical emissions, all occurring simultaneously. These are joined by the recent spread of antibiotic-resistant bacteria, microplastic nanopollutants, and the rapid increase of autism spectrum disorders, obesity, allergies, asthma, and other adverse health effects. The social stress, political chaos, and economic collapse of global consumer society that accompany these multiplying environmental threats constitute the reality of biocatastrophe as a simultaneous field of ongoing crises.

The Ritual of Evasion

Biocatastrophe as a simultaneous field of multiple ongoing interrelated ecological and humanitarian crises remains off limits for media coverage and thus for informed debate. The ritual of evasion pertaining to the interrelated nature of the multiple crises that constitute biocatastrophe reflect the fundamental reluctance of modern industrial/consumer society to face the reality of the consequences of 6,000 years of pyrotechnic activities in our vulnerable round world biosphere.

We live in the early stages of the Age of Biocatastrophe. Why are we so reluctant to document and debate the interrelationship – the synergism – of its many ongoing crises?

www.ingramcontent.com/pod-product-compliance
Lightning Source LLC
Chambersburg PA
CBHW062026210326
41519CB00060B/7184